AF694109
WITHDRAWN

# Planar Lipid Bilayers

# BIOLOGICAL TECHNIQUES

## A Series of Practical Guides to New Methods in Modern Biology

**Series Editor**
DAVID B SATTELLE

**Computer Analysis of Electrophysiological Signals**
*J Dempster* (published November 1992)
**Fluorescent and Luminescent Probes for Biological Activity**
*WT Mason* (published May 1993)

CLASSIC TITLES IN THE SERIES

**Microelectrode Methods for Intracellular Recording and Ionophoresis**
*RD Purves*
**Immunochemical Methods in Cell and Molecular Biology**
*RJ Mayer and JH Walker*

BIOLOGICAL TECHNIQUES

# Planar Lipid Bilayers

## Methods and Applications

W. HANKE

*Institut für Zoophysiologie–Biologie*
*Universität Hohenheim*
*Germany*

W.-R. SCHLUE

*Institut für Zoologie*
*Lehrstuhl für Neurophysiologie*
*Heinrich-Heine Universität Düsseldorf*
*Germany*

ACADEMIC PRESS
*Harcourt Brace & Company, Publishers*
London · San Diego · New York
Boston · Sydney · Tokyo · Toronto

TP
248.25
.B53
H36
1993

ACADEMIC PRESS LIMITED
24–28 Oval Road
London NW1 7DX

*United States Edition published by*
ACADEMIC PRESS INC.
San Diego, CA 92101

Copyright © 1993 by
ACADEMIC PRESS LIMITED

*All rights reserved*
No part of this book may be reproduced in any form, by photostat, microfilm, or by any other means, without written permission from the publishers

This book is printed on acid-free paper

A catalogue record for this book is available from the British Library

ISBN 0–12–322994–4 (hardback)
ISBN 0–12–322995–2 (bench-top edition)

Typeset by J&L Composition Ltd, Filey, North Yorkshire
Printed and bound in Great Britain at The Bath Press, Avon

# Series preface

The rate at which a particular aspect of modern biology is advancing can be gauged, to a large extent, by the range of techniques that can be applied successfully to its central questions. When a novel technique first emerges, it is only accessible to those involved in its development. As the new method starts to become more widely appreciated, and therefore adopted by scientists with a diversity of backgrounds, there is a demand for a clear, concise, authoritative volume to disseminate the essential practical details.

*Biological Techniques* is a series of volumes aimed at introducing to a wide audience the latest advances in methodology. The pitfalls and problems of new techniques are given due consideration, as are those small but vital details that are not always explicit in the methods sections of journal papers. The books will be of value to advanced researchers and graduate students seeking to learn and apply new techniques, and will be useful to teachers of advanced undergraduate courses, especially those involving practical and/or project work.

When the series first began under the editorship of Dr John E Treherne and Dr Philip H Rubery, many of the titles were in fields such as physiological monitoring, immunology, biochemistry and ecology. In recent years, most biological laboratories have been invaded by computers and a wealth of new DNA technology. This is reflected in the titles that will appear as the series is relaunched, with volumes covering topics such as computer analysis of electrophysiological signals, planar lipid bilayers, optical probes in cell and molecular biology, gene expression, and *in situ* hybridization. Titles will nevertheless continue to appear in more established fields as technical developments are made.

As leading authorities in their chosen field, authors are often surprised on being approached to write about topics that to them are second nature. It is fortunate for the rest of us that they have been persuaded to do so. I am pleased to have this opportunity to thank all authors in the series for their contributions and their excellent co-operation.

David B Sattelle ScD

To Marianne
To Ute

# Preface

Biological membranes are highly complicated systems with a very sophisticated biochemical composition. Detailed information on membrane-bound processes is fundamental to an adequate understanding of living systems in general, but is often precluded by the complicated nature of the system itself. To make a simplified model of such a system and study this before investigating the 'real world' is the tradition in experimental physics, and has been applied to biological membranes too.

In the early 1960s, a way was found to build artificial membranes from the main chemical components, lipids, and to incorporate simple transport processes, for example ion carriers and pores, into these 'planar lipid bilayers'. In what follows, we will show how the investigation of planar lipid bilayers has become an established area of research with wide applications. Transport-mediating components, proteins, have been isolated from biological membranes and incorporated into artificial membranes. There, they can be studied and their properties can be characterized in detail in a chemically defined environment. In addition, theories have been developed describing the planar lipid bilayer as a thermodynamic system.

Numerous papers have been written dealing with planar lipid bilayer experiments. Surprisingly, and in contrast to many other areas of membrane research, very few of these reports provide an extended coverage of practical details of planar lipid bilayer research. This may be due to the difficulties sometimes experienced by newcomers in the reproduction of bilayer experiments, in problems of maintaining bilayer stability and in technical problems of controlling all the variables of an artificial membrane-system.

If this volume helps to demystify bilayer experiments then it will have served its purpose.

The aims of this text are not only to assist the beginner but also to provide experienced scientists with some technical tips. Some guidelines for trouble shooting in bilayer experiments are also included. We will start from the basics of membrane physics and work towards a functional bilayer set-up and a set of experimental procedures that should make bilayer experiments less frustrating. We hope through this to minimize the difficulties of planar lipid bilayer experiments and to interest more scientists in this technique.

Following the Introduction, Chapter 2 reviews the theoretical background useful for a beginner in the field. This chapter may not be crucial for scientists with some knowledge in the field. Chapters 3–7 deal with technical questions of bilayer experiments; they form the core of this book and provide the essential information needed for successful bilayer experiments. In Chapter 8, some examples of bilayer experiments and their results are discussed, starting from simpler questions and ending with more sophisticated reconstitution experiments. Together with Chapters 3 and 7, Chapter 8 also contains plans and details of materials which are given with the aim of assisting those interested in the construction of a functional set-up. Finally, in Appendix 4 some addresses are listed for companies producing equipment, chemicals and software useful for bilayer experiments. At the end of the book a collection of original literature is given.

**Acknowledgements**

It is a pleasure for us to acknowledge here all the help we have received in learning what we know about bilayers, and in the writing of this book.

Without the time W. Hanke spent in the laboratories of Günther Boheim (Bochum, FRG) this book would not have been written.

We wish to thank H. Horn, E. Friedrich and E. Gerbershagen who helped us (in fact they gave much time for it) in the time-consuming task of preparing the artwork of the book.

We thank Academic Press and the series editor D.B. Sattelle for encouraging us to write the book.

Finally we want to thank all those who gave us permission to use examples of their work in illustrating the text.

W. Hanke
W.-R. Schlue

# Contents

CHAPTER ONE

# Introduction

## 1.1 PLANAR LIPID BILAYERS: TOOLS FOR INVESTIGATING BIOLOGICAL MEMBRANES

Biological systems are organized in cells, which may exist as unicellular or multicellular organisms. The cells are separated from their environment and from other cells by membranes. In addition to the outer membrane (plasma membrane), cells are structured and compartmentalized by systems of internal membranes. Intracellular organelles such as mitochondria or chloroplasts are surrounded by membranes. All the different membranes have to fulfil a large variety of functions, and this is reflected in their structure and composition. Looking at the structure of a cell, a few of these functions are obvious immediately.

Living organisms are critically dependent on the communication of cells with their environment and with each other. Because of the existence of the membranes, each cell is a closed system. Thus the plasma membrane must permit the cell to communicate both with its environment and with other cells. In addition, there must be pathways for transport of material across the membranes from the outside to the inside and vice versa. For this reason, the plasma membrane (and inner membrane systems, too) must be able to co-ordinate the transport of material from the outside to the inside of the cell and back. This transport must be selective, with appropriate control mechanisms to regulate it. For communication, the membrane must be able to transfer signals from one side of the membrane to the other. Additionally, lateral signal flux in the membrane should be possible, and the membrane must be electrically excitable. Finally, the membrane must enable the cell to receive signals from its environment and to export signals to it. In some cases cells are also required to release (or receive) specific chemical signals from other cells of the same or of a different organism (see Alberts *et al.*, 1983, for example).

As we are mainly interested in the membrane

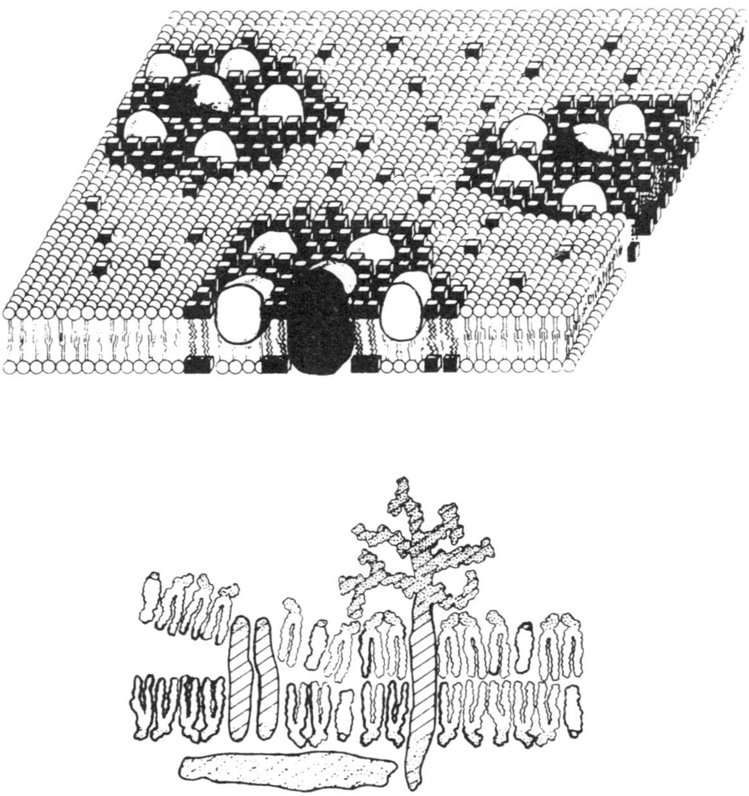

**Figure 1.1** According to Singer and Nicholson (1972), the membrane is a lipid bilayer in which some proteins are embedded (intrinsic) and with which other proteins may be associated (extrinsic). Unless fixed, as is the case for associated proteins linked to for example the cytoskeleton, membrane proteins are free to diffuse within the membrane. Additional components may be added to the system, making it more complicated. Upper: two-dimensional representation of a membrane with a hypothetical domain structure. Lower: linear model with proteins (hatched) embedded in or associated with the lipid core. (Reproduced and modified from Hoppe *et al.*, 1982 with permission.)

itself, and in practice not in the real membrane but in simplified models, it is inappropriate to provide an extensive review of membrane structure and function here. The interested reader is referred to textbooks on cell- and membrane-physiology (see Bretscher, 1985; Finean, 1984; Appendix 5).

The structure of biological membranes which is necessary to fulfil all the functions stated above has been investigated extensively and some common consensus has now been reached. Singer and Nicholson (1972) established the so-called 'fluid mosaic model' of biological membranes which, with some modifications, still provides the basis of current membrane models. Figure 1.1 shows a simplified schematic representation of elements of this model. The model envisages a double layer of lipid molecules, the hydrophilic head-groups of which are exposed to the aqueous phases. The hydrophobic hydrocarbon chains of the lipid molecules are oriented towards each other yielding the hydrophobic membrane core. Lipid double-layer proteins (and other molecules) are associated with this. Additionally, in the 'lipid bilayer' there are proteins spanning the entire membrane (integral, or intrinsic, proteins) (associated, or extrinsic, proteins) only partly embedded. Such membranes may be symmetrical as well as asymmetrical in construction. To our present knowledge, biological membranes are always asymmetrical, i.e. the two monolayers from which the bilayer is built differ in their composition in both lipid and protein content. The proteins embedded in the lipid bilayer may freely diffuse in the membrane, alternatively they may be fixed to a certain location in the cell membrane by structures inside the cell, which are not shown in Figure 1.1. Additionally, as shown in Figure 1.1 membranes are not necessarily homogeneous but may be structured in domains of different composition, making them even more complex.

As the lipophilic region of bilayer membranes is highly hydrophobic it is evident that transport (by diffusion) of ions across the bilayer itself is not possible at rates observed *in situ*. Furthermore, transport simply by diffusion or any other uncontrolled passive mechanism through a pure lipid membrane can hardly fulfil the prerequisites of regulation, selectivity and control of flux for any given material. Obviously the proteins are the only entities that can support the necessary pathways across the membranes, including for example hydrophilic pores for ions. It has been established that proteins are the molecular basis for membrane excitability and that they receive and transduce signals from the external world into appropriate cellular responses.

However, it is not simply the case that membrane proteins are embedded in a lipid bilayer doing their job and otherwise not interacting, the bilayer simply serving as a mechanical support. The lipid composition of a membrane can profoundly change the protein-mediated membrane processes, and physical parameters can alter the whole system. Such physical parameters can affect the thermodynamic state of the membrane, which can be considered as a thermodynamic multi-component, multi-phase system (Hoppe *et al.*, 1982).

The functional diversity of membranes is reflected in the variety of proteins and lipids that

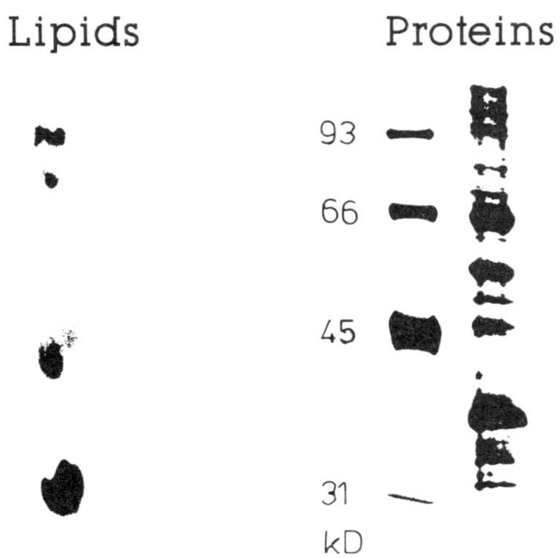


**Figure 1.2** Lipid and protein composition of a membrane fraction. A thin-layer chromatograph of soya bean lipids is shown together with an SDS-polyacrylamide gel of the protein extract of a biological membrane. For convenience markers for protein size are included.

can be detected in them. These components make up biological membranes, the final composition of which can be extremely complex. Methods have been developed to study the composition of membranes, and results of such an experiment are shown in Figure 1.2. An SDS-gel of part of the protein moiety of a membrane is presented together with thin-layer chromatography of its lipids. It is clear that a large number of components are present, even though not all are depicted in the example shown. The study of these complicated systems raises some problems. First, they cannot be described by complete physical models because the still limited capacity of modern computers is not adequate; and acceptable theories for such complicated systems are at present not available. Second, when describing isolated parts of the whole system, each may be affected by uncharacterized parts of the *in situ* membrane, so it is impossible to exclude interactions of different membrane regions and components. One approach to solving such an unacceptable situation is to simplify the system. A reduced model–system must be constructed which enables manipulation of all components under controlled conditions. After having understood the simple model-system(s), further complexity can be introduced, different models can be combined and, finally, one may achieve the description of a biological membrane or something close to it. A strategy of this kind is summarized in Figure 1.3.

The most simple membrane model is a planar lipid bilayer composed of one pure lipid, and this is the first system one should construct, investigate and understand: Figure 1.4 illustrates the major elements of such a system. It is necessary to state here that, to date, even this aim has not been finally achieved. The next step would be to include one more component, let us say a protein, creating an aqueous pore across the bilayer. Now we have a two-component system which is still very simple (compared to a real biological membrane). In this way, the properties of at least the transport system might be understood. The composition of such a hypothetical two-component system is shown in Figure 1.5. Later the system can be made more and more complicated by adding other

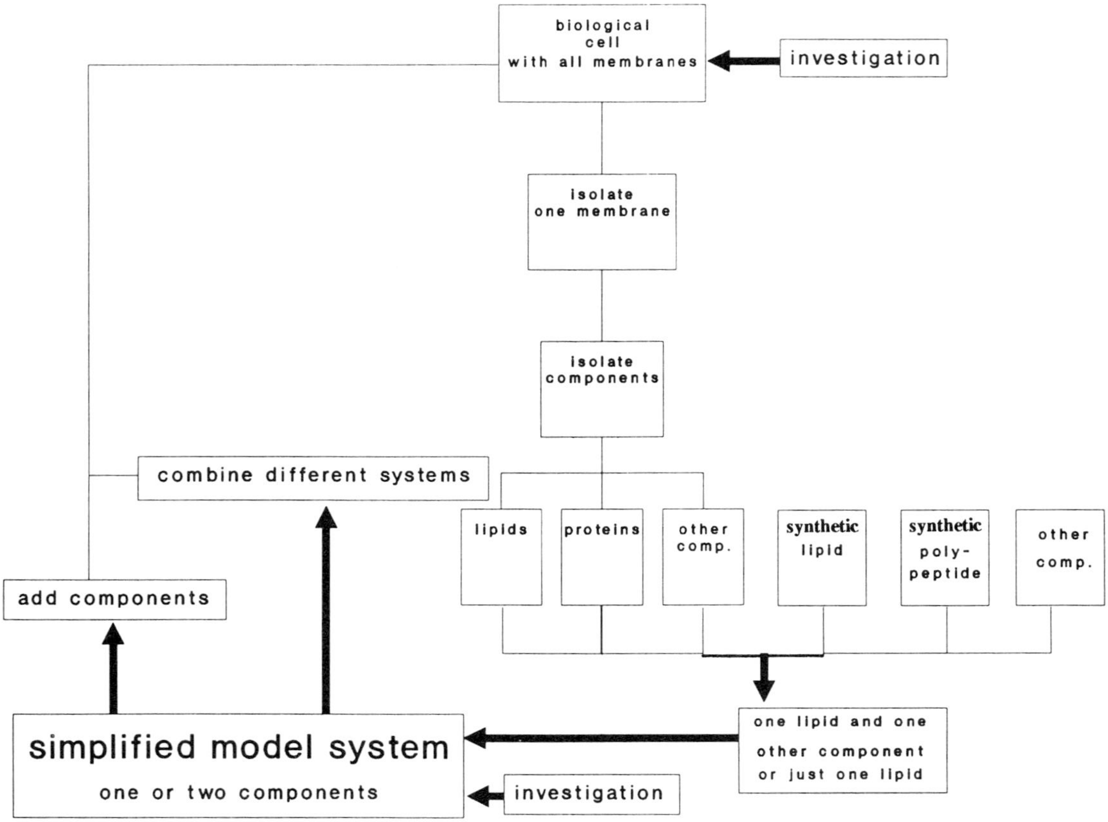


**Figure 1.3** Schematic representation of a strategy appropriate for the investigation of membrane properties. By these means a natural system is reduced to a simplified, artificial system. This provides a well-defined starting point from which it is possible to investigate a biological membrane step by step.

components or by combining data from different isolated models.

After having identified such a strategy, and of course others are possible, (we will come back to this question later, when discussing the patch-clamp technique), one has to solve the problem of how to achieve this most effectively. Following the ideas presented above, it is obvious that it would be worth constructing a 'planar lipid bilayer', or a curved shell of it, from let us

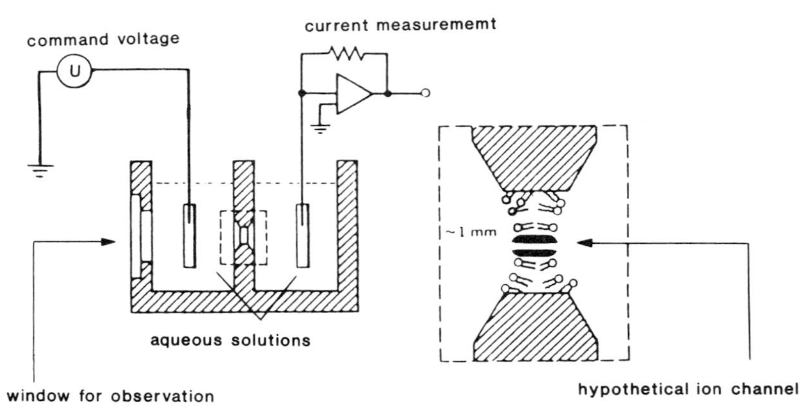


**Figure 1.4** The principles of a bilayer experiment. An artifical membrane is constructed over an aperture in a partition between two aqueous phases. Ease of access to both sides facilitates direct electrical measurement and manipulation of composition of the bathing solutions. The basic circuit for measurement of the current flowing through a hypothetical channel is shown.

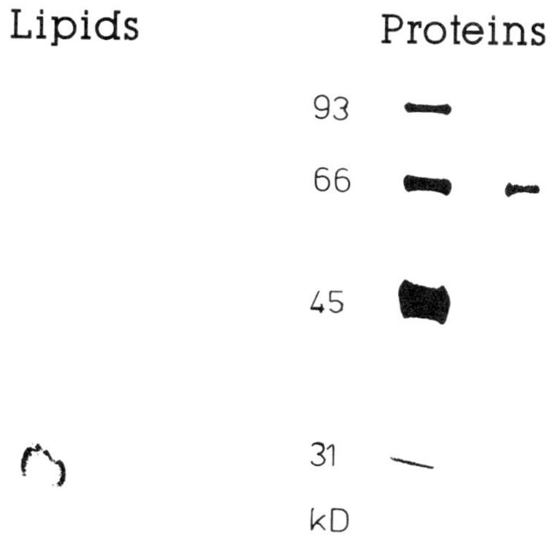


**Figure 1.5** Using data of the type shown in Figure 1.2 the composition of an idealized artificial membrane is shown, made from a synthetic lipid (thin-layer chromatography) and a purified protein (SDS-gel). A planar lipid bilayer (made from one synthetic lipid) with one channel-forming polypeptide incorporated can be such a system.

say a synthetic lipid, studying this and then incorporating purified proteins into it and studying them (Mueller *et al.* 1962; Montal and Mueller, 1972; Miller, 1986). It is the aim of this book to provide the theoretical and practical basis for this and to deal with the interpretation of results from such experiments.

To give a general idea of how bilayer experiments can contribute to our knowledge of ion channels Figure 1.6 depicts ways to investigate one type of ion channel using different methods. Of course, bilayer experiments are not sufficient for discovering everything about an ion channel, they are just one of several possible approaches, and an ion channel is only one isolated component from a biological membrane. In Figure 1.7 an expanded scheme shows the bilayer aspects of such a study.

Other techniques also have been developed for investigating isolated processes from biological

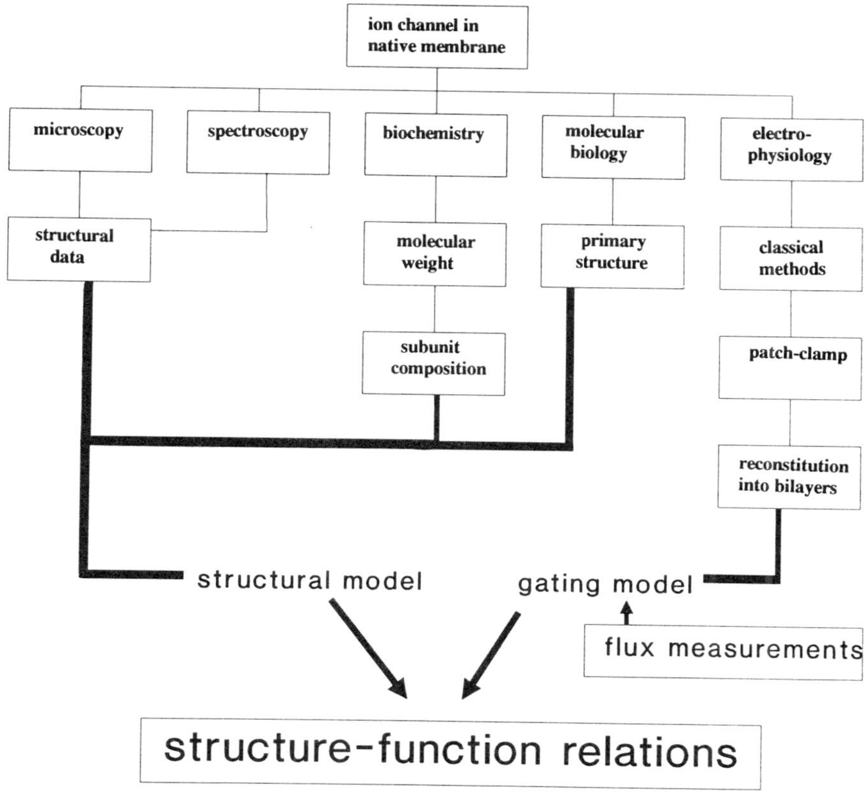


**Figure 1.6** Strategy for the investigation of ion channels from biological membranes. Different methods are used to investigate ion-channel properties. The final aim here is to understand the relations between molecular structure and function of the protein. Of course, other questions may be asked as well, when using the techniques shown.

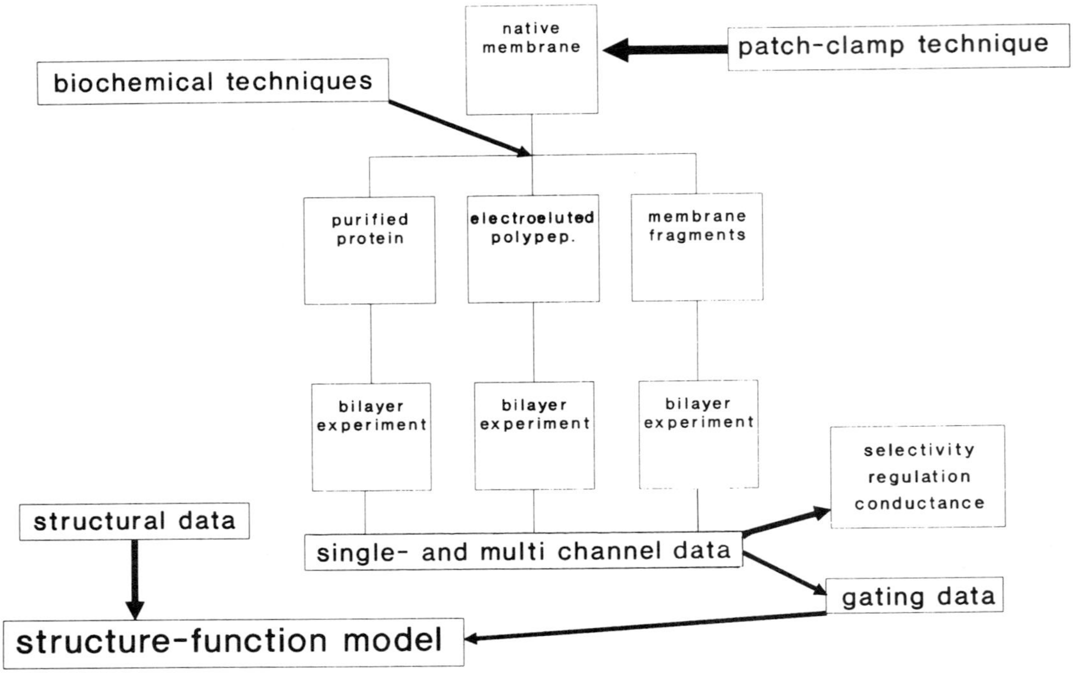


**Figure 1.7** The bilayer section from Figure 1.6 is here expanded to show that even with such simplified models many different investigations are possible. Obviously, much information can be derived from such bilayer experiments with respect to understanding the detailed functions of ion channel proteins.

membranes. All these techniques are to some extent reductionist in character, to simplify the complicated structure of biological membranes. The only other technique besides planar lipid bilayer experiments, allowing electrical recording of isolated (single) ion channel fluctuations is the patch-clamp technique (Hamil *et al.*, 1981). Therefore we will discuss it here, though in outline only, pointing out its own unique contribution to membrane research.

## 1.2 THE PATCH-CLAMP TECHNIQUE

As stated above, biological membranes are highly complicated in their composition. The idea of the patch-clamp technique (Sakmann and Neher, 1983) is to isolate a very small area of a biological membrane and investigate it in detail. When this area is made small enough it may contain just one functional transport protein, which can then be investigated with electrophysiological techniques in isolation from the remaining (complicated) part of the membrane.

As shown in Figure 1.8 (upper) a glass pipette with a tip diameter in the micrometre range can be attached to a membrane. By application of slight suction (to the inside of the pipette), a very tight seal between glass and lipid can be achieved (in the giga-ohm range) and the membrane region under the pipette can be electrically isolated from the rest of the plasmalemma. After establishing the seal, the pipette can be moved away from the cell (Figure 1.8 lower) and a 'cell-free' patch can be established, providing good access to one side of the patch. This cell-free membrane 'patch' can be investigated in detail using electrophysiological techniques.

The patch-clamp technique has contributed a lot to our present knowledge of membrane physiology, which is reflected by the enormous literature on this topic. The interested reader is refered to the literature for detailed information (see Appendix 5).

When compared with artificial systems (planar

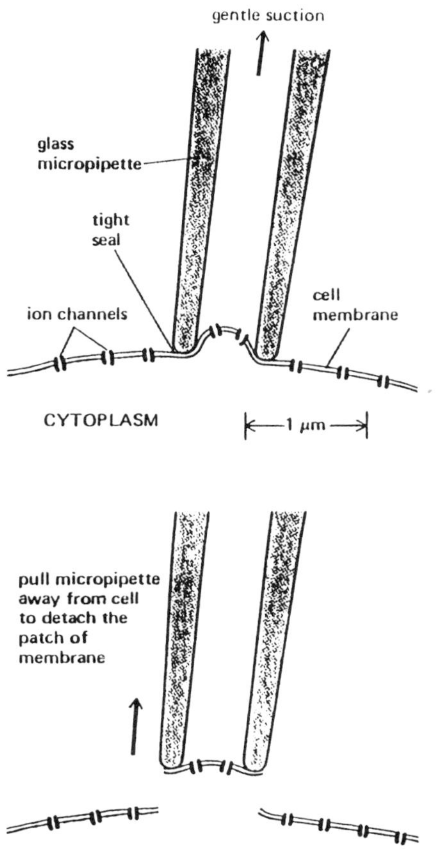


**Figure 1.8** Cartoon of the patch-clamp technique. A small area of a biological membrane is isolated electrically under the tip of a glass micropipette (upper). Membrane parameters and functional proteins (in the ideal case one molecule) in this small area can be investigated in detail, in isolation from the remaining part of the cell membrane. When the pipette is moved away from the cell (lower), a cell-free patch can be established with good access to one side of the membrane (Alberts *et al.*, 1983).

lipid bilayers), the patch-clamp technique does have some advantages:

1. it can be applied directly to biological membranes
2. it is easier to handle, as no biochemical isolation of lipids or proteins is required and no membrane needs to be rebuilt
3. usually patch-clamp experiments are more reproducible and more stable than bilayer experiments
4. the electrical resolution is very high.

On the other hand some disadvantages of the patch-clamp technique are evident in such comparisons:

1. the membrane composition cannot be controlled
2. access to the precise physical state of the membrane is not available
3. it is difficult to modify the solution bathing the face of the membrane exposed to the pipette's interior during experiments
4. not all membranes can be reached by a glass pipette
5. finally, even a small membrane patch is usually very complicated in its composition and cannot be described completely in thermodynamic terms.

There are other advantages and disadvantages to the patch-clamp and the bilayer techniques, but the above comparison should serve to demonstrate that care is needed to select the approach better suited to the problem under investigation (see Figures 1.3 and 1.6).

More recently, techniques have been developed that attempt to combine the advantages of both the bilayer and the patch-clamp techniques; notably the technique of making bilayers on the tip of glass pipettes. These techniques, which are referred to as 'dip-stick', or simply 'dipping', techniques are discussed in the following chapters.

## REFERENCES

Alberts, B., Bray, D., Lewis, J., Raff, M., Roberts, K. and Watson, J.D. (1983) *Molecular biology of the cell.* Garland Pub. Inc., New York.

Bretscher, M.S. (1985) The molecules of the cell membrane. *Sci. Am.* **253**: 100–109.

Finean, J.B., Coleman, R. and Michel, R.H. (1984) *Membranes and their cellular function.* Blackwell Scientific Publications.

Hamil, O.P., Marty, A., Neher, E., Sakmann, B. and Sigworth, F. (1981) Improved patch-clamp techniques for high resolution current recording from cells and cell-free patches. *Pflügers Arch.* **391**, 85–100.

Hoppe, W., Lohmann, W., Makl, H. and Ziegler, H. (1982) *Biophysik*. Springer, Berlin.

Miller. C. (1986) *Ion channel reconstitution.* Plenum Press, New York.

Montal, M. and Mueller, P. (1972) Formation of

bimolecular membranes from lipid monolayers and a study of their electrical properties. *Proc. Natl. Acad. Sci. USA* **69**, 3561–3566.

Mueller, P., Rudin, D., Tien, H.T. and Wescott, W.C. (1962) Reconstitution of excitable cell membrane structure in vitro. *Circulation* **26**, 1167–1171.

Neher, E. and Sakmann, B. (1983) *Single channel recording*. Plenum Press, New York.

Singer, S.J. and Nicholson, G.L. (1972) The fluid mosaic model of the structure of cell membranes. *Science* **175**, 720–731.

CHAPTER TWO

# Physical properties of biological membranes and planar lipid bilayers

At their simplest, biological membranes are hydrophobic barriers in a hydrophilic environment. However, in living cells they have to fulfil a variety of functions to guarantee their viability in a changing environment. A major task of membranes is to enable controlled transport across the otherwise impermeable membrane. Selectivity of these transport properties is mainly achieved by proteins incorporated into the lipid core of the membranes (Findlay and Evans, 1987; Höfer, 1977). In this chapter we will discuss a few basic theoretical features of pure lipid membranes and protein-mediated transport across such barriers, mainly in the context of the requirements for planar lipid bilayer experiments.

Before proceeding later to some details about real membranes we must first make a few statements about pure lipid membranes. At the outset we will therefore consider a few very basic theoretical points.

## 2.1 PHYSICAL PROPERTIES OF PURE LIPID MEMBRANES

Lipids are amphiphilic molecules; they have hydrophilic and hydrophobic regions. As a consequence, on contact with an aqueous solution they will tend to adopt particular well-defined orientations. At a water–air interface, lipids will form structured monolayers as shown in Figure 2.1 (upper). In this figure the structure of a typical lipid molecule, a phospholipid, is also shown. A monolayer as shown in Figure 2.1 (upper) is a two-dimensional thermodynamic system, for which an adequate theoretical description can be provided. A good deal of information gathered in the investigation of lipid monolayers at air–water interfaces is also valid for planar lipid bilayers. Data on phase transitions, as well as information on lipid packing density, are measured in monolayer set-ups often called film balances (Figure 2.1 lower left). A typical measurement with such a film balance is depicted

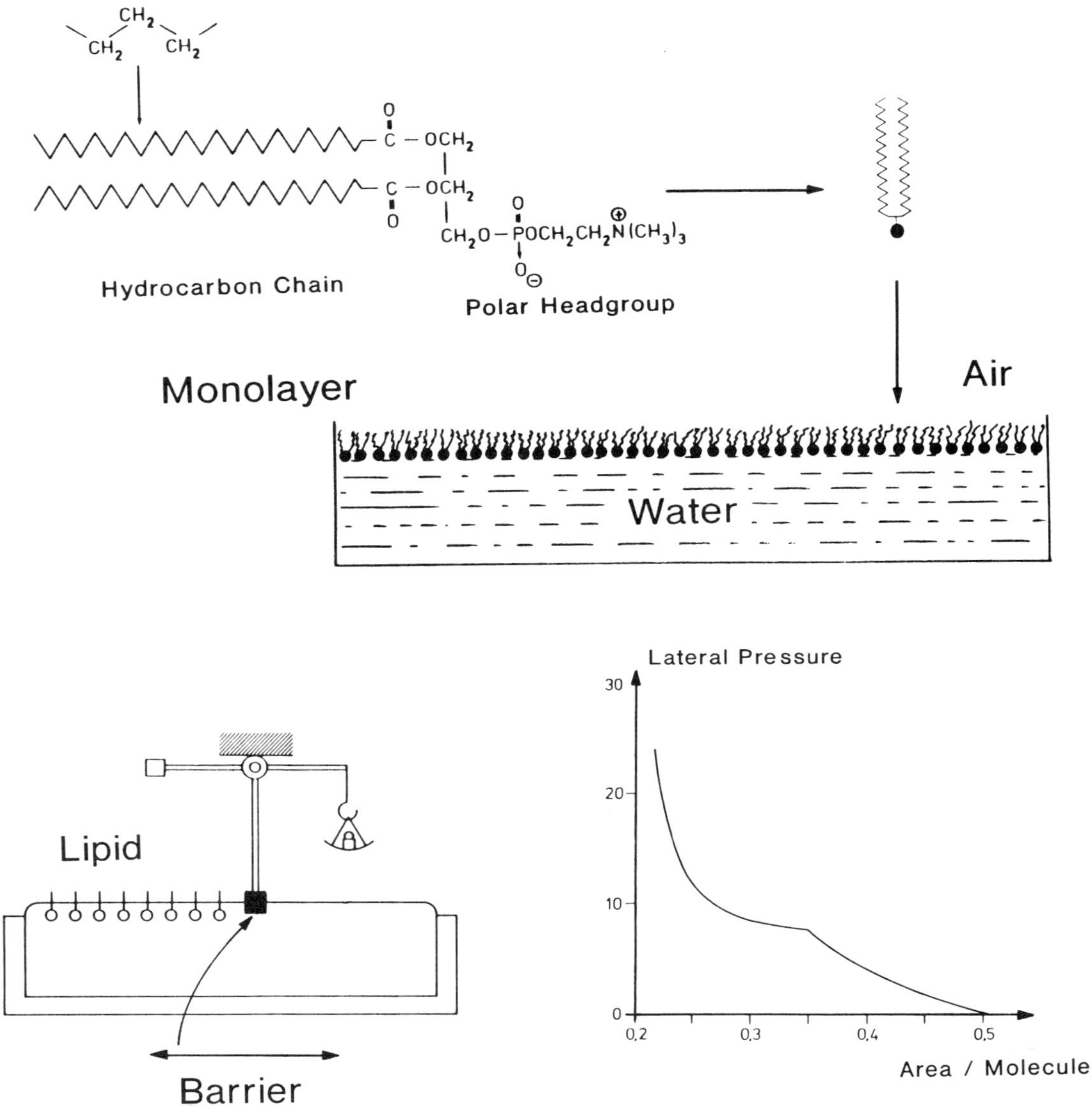


**Figure 2.1** Upper: In the upper left part of the figure the structure of a phosphatidylcholine (lecithin) molecule is given as an example of a lipid-structure. It consists of the hydrophilic head-group, here choline, the glycerol skeleton and the two fatty acids forming the hydrophobic part of the molecule. At an air–water interface these amphiphilic molecules build structured monolayers as shown. The fatty acids are oriented towards the air, the hydrophilic choline (or other) head-groups are oriented towards the aqueous phase. Lower: Monolayers (lipid films) can be investigated as two-dimensional thermodynamic systems with so-called film balances. The main parts of a film balance are a trough for the aqueous phase on which the lipid film is spread, a barrier which is used to change the area of the lipid film and a device to measure the surface tension of the film. With such film balances the thermodynamics of lipid films have been studied in detail with respect to a variety of parameters. In addition to the diagram of the film balance, the protocol of a typical experiment is shown. Lateral pressure in the lipid monolayer is shown as a function of area per lipid molecule. The example given exhibits a first-order phase transition. The graph axes are scaled in arbitrary units (Reproduced and modified from Adam *et al.*, 1977 with permission).

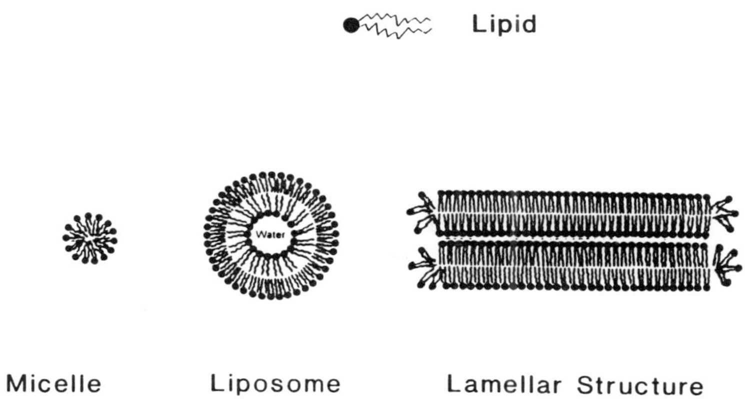


**Figure 2.2** In contact with water, lipid will form certain defined geometrical structures due to its amphipathic structure. The final geometry depends on the lipid and its concentration in the water. Independent of these details a few possible structures are shown. They all have in common that the hydrophilic head-groups are oriented towards the aqueous phase and the fatty acids (or hydrophobic areas) are oriented towards each other.

in Figure 2.1 (lower right). Here a graph of the lateral pressure in the monolayer as a function of area per lipid molecule is given; the example shown has a first-order phase transition. Lipid monolayers in fact provide the starting point for some important new methods of constructing bilayers. Some details on this aspect of lipid monolayers will be given later, but for details on the wider application of monolayers the reader is referred also to specialized literature (see Starzak, 1984).

In contact with aqueous solutions, other structures can also be formed by lipids. Due to the amphiphilic structure of the lipid molecules, they can aggregate with their hydrophilic parts oriented towards each other. The structures which are formed in these lipid-water systems are dependent on the structure of the lipid and on the concentration of the lipid in the solution. A collection of possible geometries fulfilling the physical constraints of energy minimization are given in Figure 2.2 (independent of structural and concentration considerations).

We can now, for reasons of simplicity, define pure lipid membranes (planar lipid bilayers without any additional component) as lipid double layers in a hydrophilic environment which may be flat or have some curvature. A sketch of such a system is given in Figure 2.3.

A very basic electrical equivalent of the same system is a circuit of a parallel capacitor and a resistor (Figure 2.3, upper). The resistor reflects the pure electrical conductance of the bilayer. The conductance of the pure lipid membrane is very low, but not negligible in experiments with high current resolution. In a circuit more realistically equivalent to the simple bilayer system itself, at least the resistors made up by the electrodes and aqueous phases must be added.

In a standard experiment a membrane potential is applied across such a membrane. At a constant potential this gives a constant (very low) current across the membrane. Pure lipid membranes respond in a manner comparable to ohmic resistors when physiological potentials are applied. However, at higher potentials ($> [\pm\ 150\ mV]$) there is often a significant increase in conductance, usually due to electrical membrane breakdown and the opening of unspecific (lipid) pores in the lipid bilayer (Benz *et al.*, 1979). This effect has been used for example in the technique of electroporation to permeabilize cell membranes (Benz and Zimmermann, 1980), for example to

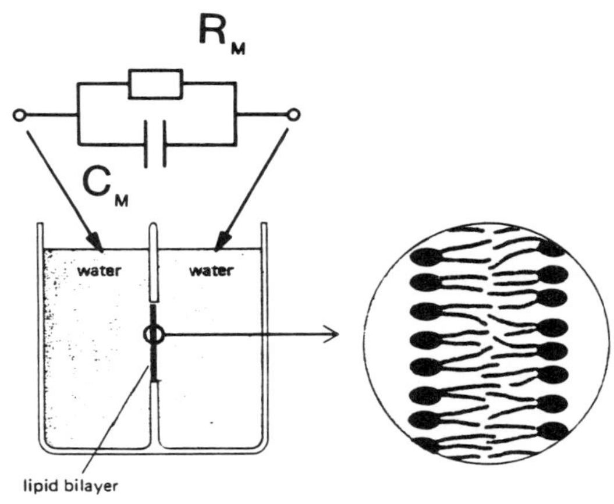


**Figure 2.3** Sketch of a planar lipid bilayer together with the most simple electrical equivalent circuit of such a bilayer, consisting of a resistor ($R_m$) and a capacitor ($C_m$) in parallel.

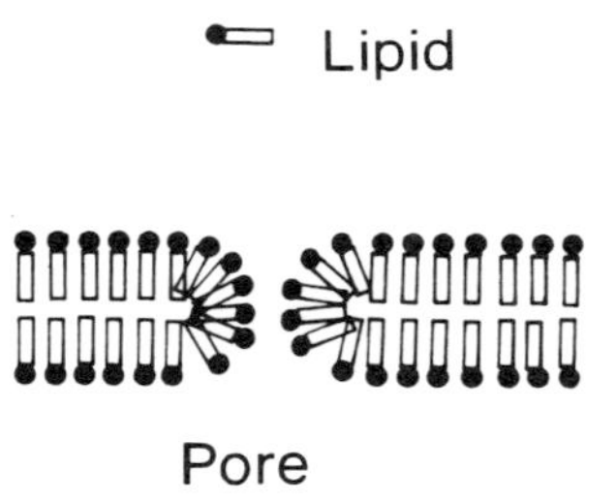


**Figure 2.4** Structure of an aqueous pore in a pure lipid membrane. Due to the energetics, only the hydrophilic head-groups of the lipid molecules can be exposed to the water phase, thus the structure shown is the only possibility for such a pore because the water filled tunnel in the centre must be lined by the hydrophilic head-groups of lipid.

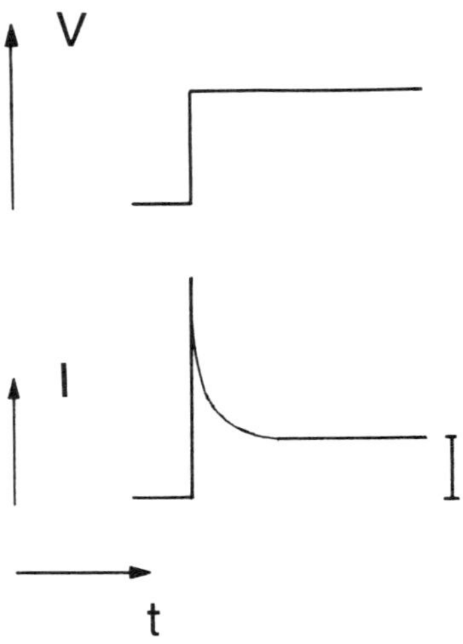


**Figure 2.5** Response of a bilayer to a stepwise increase of membrane potential, the command potential protocol (V) and the current response (I) are given. Due to the capacity of the bilayer, there is first a transient current spike when the capacity is loaded; this relaxes to a current value defined by the ohmic resistance of the bilayer. The peak value of the spike is limited by the external resistors of the system (electrodes, potential source, etc.).

transfect them with exogenous DNA (Potter *et al.*, 1984). The possible structure of a hydrophilic pore in a pure lipid membrane is shown in Figure 2.4. In real bilayer experiments, such pores are responsible for the non-ohmic part of the *I*/*V* relation of pure bilayers at high potentials and they often induce membrane instability, which can eventually lead to the breakdown of the entire membrane.

When a potential which varies in time is applied across a bilayer, the capacitative current across the bilayer must be taken into account. The capacitance of a bilayer is the result of a non-conducting medium, the lipid, being surrounded by conducting aqueous medium, and is therefore comparable to a simple plate capacitor as known from basic electrodynamics. The current across such a capacitor is given by:

$$I = C \times \mathrm{d}V/\mathrm{d}t \qquad [2.1]$$

where $I$ = current across the capacitor, $C$ = capacitance. The capacity of the membrane capacitor is given by

$$C = \varepsilon_o \varepsilon_r + A/d \qquad [2.2]$$

where $\varepsilon_o$ = absolute dielectric constant, $\varepsilon_r$ = relative dielectric constant of the membrane (about 2–3 in most cases, compared to 80 for water), $A$ = membrane area and $d$ = membrane thickness.

A typical simple experiment with a planar lipid bilayer is the application of a step-like potential change to it. The protocol for such an experiment is given in Figure 2.5. Here following the potential step, first the capacitance is charged, resulting in a spike in the current trace, which subsequently decreases to a steady state value determined by the resistance of the membrane.

Figure 2.6 summarizes the description of the lipid bilayer in terms of a simple electrical equivalent circuit. The membrane is represented by a parallel resistor and capacitor, as seen already in Figure 2.3. Additional resistors represent the electrodes and the aqueous solutions, and a voltage-dependent resistor which represents the non-ohmic behaviour of pure lipid membranes, especially at higher potentials, is added. A potential source and a current measurement device are included to yield a more practical system.

For bilayers composed of neutral lipids, this simple description as an electrical equivalent circuit is sufficient in most bilayer experiments. The electrical field across such a membrane in physiological salt solution can be described by a simple model (Figure 2.7, upper). The situation becomes more complex when charged compounds (usually negatively charged lipids, which exist in

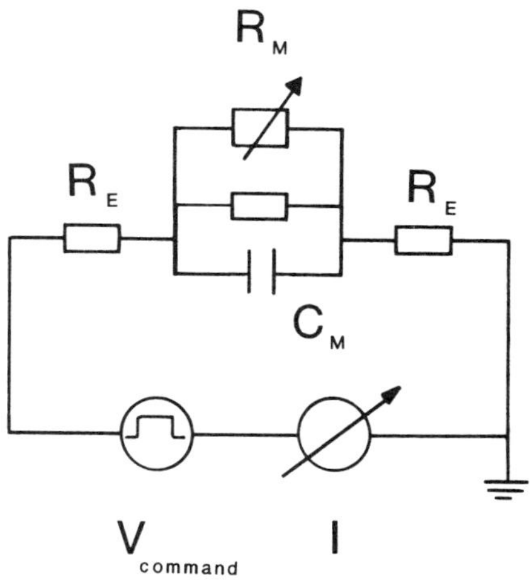


**Figure 2.6** Electrical equivalent circuit of a planar lipid bilayer in a real set-up. It contains the bilayer itself, with membrane resistance ($R_M$), membrane capacity ($C_M$) and a voltage dependent-resistance ($R_E$), the electrodes and aqueous solutions, the potential source ($V_{command}$) and a current measurement device (I).

nearly all biological membranes) are constituents of the bilayer. The presence of such compounds leads to a situation as given in Figure 2.7, (centre), other possible potentials, i.e. from dipole moments, are included in the figure. The membrane will carry a surface charge and the ions around the membrane will no longer be homogeneously distributed; the positive ions will partially compensate the charge density of the membrane surface and the field will appear as shown in Figure 2.7 (lower). Potential-dependent processes within the membrane may be strongly influenced by this effect. A detailed description of membrane-surface charges is given in many textbooks on membrane biophysics (e.g. Adam *et al.*, 1977).

Due to the hydrophobic nature of the lipid core of a membrane, the free diffusion of ions across such a bilayer is severely restricted. The core of such a membrane, which is about 3–5 nm thick depending on the biochemical composition, can best be described theoretically in terms of a very high energy barrier, *E* for ions. Using standard values for ion size and dielectric constant values for *E* of the order of 50–60 kcal/mol can be calculated. This high value reduces

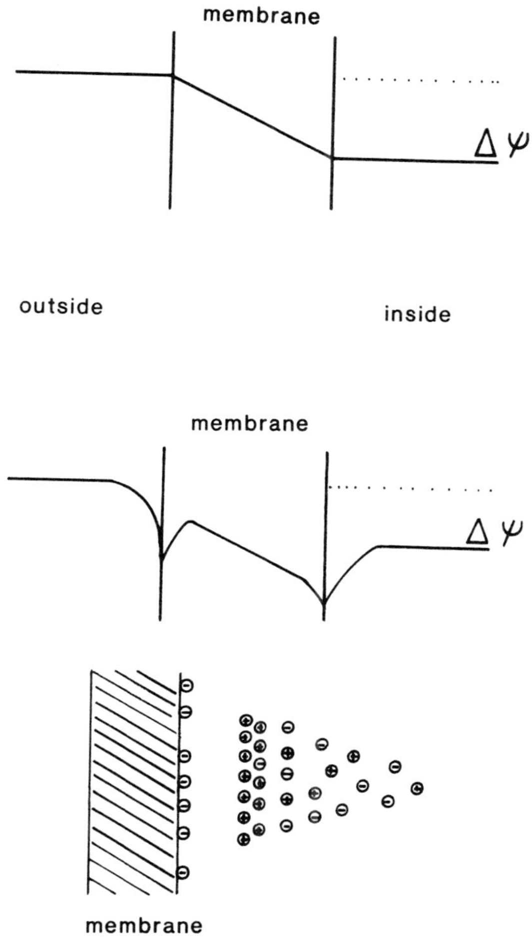


**Figure 2.7** Upper: The drop in the electrical field across a membrane composed from neutral lipids is shown. Centre: The drop in the electrical field across a membrane containing negatively charged lipids is given. The consequence of the charged lipids is a surface charge which effects the distribution of ions around the membrane. This leads to a change in the potential profile along the system as shown. Lower: The distribution of ions close to a negatively charged membrane is given.

the probability of an ion crossing the membrane about $10^{-40}$-fold compared to a neutral particle of identical size ( Parsegian, 1969; Hille, 1975). From this low probability of ions crossing pure lipid membranes, it is obvious that the electrical conductance of a pure lipid bilayer cannot be very high. Experiments with pure lipid planar lipid bilayers yield values of about $10^7$–$10^8$ ohm cm$^2$. This is not comparable to the

the specific resistance of biological membranes, which ranges from $10^4$ ohm $cm^2$ down to only a few ohm $cm^2$.

Clearly, pure lipid membranes cannot fulfil the ion-transport roles of biological membranes on energetic grounds. As stated earlier, integral membrane proteins are found to be responsible for the transport of ions across biological membranes. Different structural entities, mainly proteins but other molecules as well, have been shown capable of providing transport of ions across membranes. We will discuss some basic properties of these later in the book. Nevertheless, not only ions but also neutral molecules must be transported across membranes. The energy barrier of the membrane for these molecules is much lower than for ions, so part of this transport can be via passive diffusion, following concentration gradients. However, many neutral molecules must be transported against concentration gradients and, again, molecular entities are needed to mediate this transport, thereby expending energy. In addition, the passive diffusion rates across pure lipid membranes are not adequate for the cell, because they cannot be regulated, and somehow the process must be facilitated, again by membrane proteins.

## 2.2 PROTEINS IN LIPID MEMBRANES

With the exception of passive diffusion of neutral molecules across membranes, all transport of ions and small molecules across biological membranes is due to the presence of integral membrane proteins (or polypeptides). In fact some proteins even enable the transport of quite large proteins across a membrane. We stated already in the introduction, that all such proteins are arranged in the membrane in certain geometrical structures and may either diffuse laterally within the membrane or be anchored in defined locations in the membranes by intracellular structures.

Of course, there are integral proteins other than transport proteins present in biological membranes. Their presence in the membrane will necessarily depend on the function of the particular membrane. For example, proteins may be present in a membrane to enhance its stability or to force the membrane to adopt a certain geometrical structure (curvature). Many examples of such membrane proteins have been described in detail in the literature. We will later discuss another group of membrane proteins which was found to be necessary to enable membrane fusion under certain conditions. Though a rich diversity of membrane proteins exist, the investigation of their function in planar lipid bilayers (under controlled physical and chemical conditions) is nevertheless still in its infancy.

The structure of proteins in biological membranes has been an open question for a long time. In recent years some progress has been made with the determination of the three-dimensional structure of some major integral membrane proteins. The data for this have come almost exclusively from spectroscopic experiments and X-ray scattering experiments. Besides the fact that to date the molecular structures of only a few membrane proteins have been explored, almost nothing is known about the more interesting question of the structure–function relations of such proteins. Here electrophysiological experiments (and others), preferably done with artificial membranes of well-defined biochemical composition, may deliver the functional data necessary to advance our knowledge. The structural data must come mainly from other techniques as stated above indicating that most present areas of interest in this research field must be covered by a wide selection of different technical approaches.

As our experience is mostly related to the transport properties of membranes *in situ* and the investigation and the reconstitution of transport properties *in vitro* in artificial membranes (this reflects to a certain extent the interests of the authors), we largely confine our introduction to membrane transport and its molecular basis.

## 2.3 TRANSPORT ACROSS MEMBRANES

We have stated that transport across biological membranes is mainly attributable to the existence of membrane transport proteins. To enable transport across a membrane in a biologically useful manner, such a protein should fulfil several requirements:

1. there must be mechanisms to control the transport rate
2. the transport should display a certain degree of selectivity
3. in certain cases the transport must be active, operating against an existing electrical or chemical gradient.

These properties must be reflected in the structure of the proteins involved (and in some cases by the structure of the entire membrane). In cases of transport against a gradient, energy must be provided from the metabolism of the cell to fuel the active transport.

Transport of matter in any system can be described by thermodynamic theories which are independent of mechanistic models. Knowledge of this thermodynamic basis is necessary to understand membrane transport, especially as even now little is known about the molecular basis of transport. Nevertheless, we do want to discuss here some basic mechanistic requirements of transport processes.

### 2.3.1 Neutral molecules

To some extent, neutral molecules can pass the hydrophobic barrier of a membrane. In such cases the 'transport' of such molecules across a membrane can be easily described by the laws of diffusion (see Einstein, 1956; Adam *et al.*, 1977). The flux of material will consequently follow only the existing gradients of the component under investigation. The diffusion in such a system compared to the diffusion in a pure aqueous solution is of course restricted by the diffusion barrier of the membrane.

There may be several reasons why the simple diffusion of neutral molecules across biological membranes is inadequate to fulfil the requirements of a cell:

1. the transport rate may simply be too small
2. a gradient may be unfavourable
3. pure diffusion can not (except in the most limited way) be controlled by the cell.

In such cases, protein-(or membrane-) controlled mechanisms are necessary. Several classes of functionally distinct mechanisms have been evolved by nature to do this. Most of them resemble the structures known from the transport of ions across membranes, e.g. pores and channels, carriers and active transporters. These different molecular mechanisms are all discussed in the following chapter in more general terms.

It should be noted that the transport of neutral molecules (and also ions, to a limited degree) across membranes may be regulated by changing the properties of the entire membrane. This can be done by altering the biochemical composition, a possibility which has in fact been found in biological membranes. Because of the time needed to change a membrane's composition, however, this can only be useful in processes with a long time scale. Another possibility may be to change the thermodynamic state of a membrane. One possibility may be to produce a phase transition in the membrane; planar lipid bilayer experiments have been performed demonstrating that this is possible (Boheim *et al.*, 1980), but not very much is known about such processes in biological membranes of living organisms.

Finally, consideration should be given to the processes of exocytosis (Lindau and Gomperts, 1991) and endocytosis (Trowbridge, 1991), which are widely used in biological organisms. Exocytosis is one way to circumvent the isolation of cells from their environment by membranes. Material which has to be moved out of the cell is sequestered in a vesicle inside the cell. By fusion of the vesicle with the cell membrane, its contents are emptied into the extracellular space. Signal transmission at chemical synapses, for example, appears to include exocytosis as, by this process, the chemical transmitter is delivered to the synaptic cleft. Endocytosis is in principle the opposite process: the uptake of material via formation of a vesicle by the membrane. Single cells such as protozoa and macrophages, take up a significant part of their food by endocytosis. Furthermore endocytosis in the form of pinocytosis is used to transport membrane material of a cell into the cell itself.

Exocytosis is functionally identical to membrane fusion, a process we will discuss later in

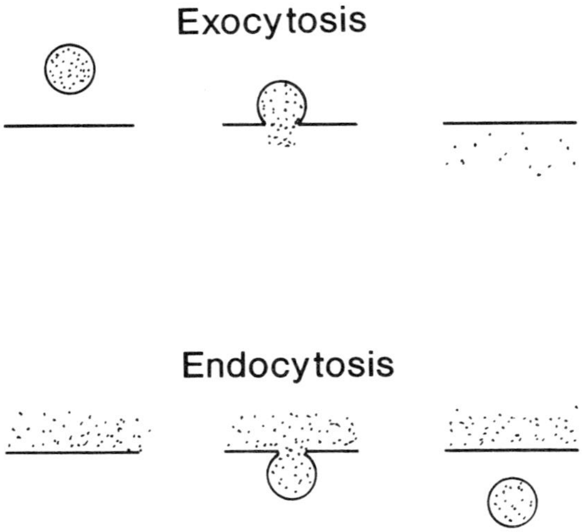


**Figure 2.8** Principles of exocytosis and endocytosis. During exocytosis (upper part of the figure) a lipid vesicle fuses with the membrane (usually inside the cell) emptying its contents on the other side of the membrane. This process is widely used, for example, in chemical synapses. During endocytosis (lower part of the figure) part of the membrane forms a vesicle which encloses a small volume of the extracellular space. The material so enclosed is then transported into the cell by separating the vesicle from the interior side of the membrane.

detail. Both endocytosis and exocytosis are represented schematically in Figure 2.8.

### 2.3.2 Ions

Ions can not diffuse across a membrane at biologically significant rates. Thus, transport processes as stated above are necessary to facilitate ion transport across membranes. Most ion transport processes which are controlled by membrane proteins are:

1. pores, ionophores and ion channels
2. carriers
3. active transporters.

This classification is of course arbitrary, but it is used widely in the literature. We will discuss pores, ionophores and channels together as they have some common mechanistic and functional principles. The pore concept additionally is often used for synthetic polypeptides and the

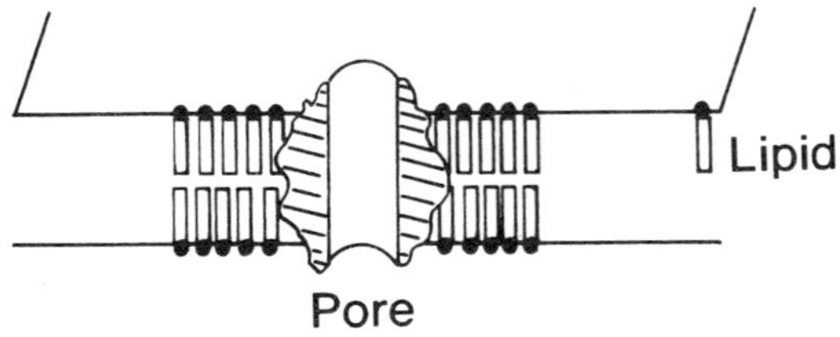


**Figure 2.9** Structure of a pore formed in a membrane by a protein. The protein, in the form of a shell or envelope, separates the hydrophilic core of the pore from the hydrophobic core of the lipid bilayer membrane, thus the protein region facing the lipids must be hydrophobic and that facing the core of the channel must be hydrophilic.

ion-channel concept is mainly used in conjunction with membrane proteins.

Systems used for the transport of ions across membranes may be used of course to transport neutral molecules, as stated in Section 2.3.1.

#### *2.3.2.1 Ion channels*

The most simple way to enable ion transport across a membrane is to create some kind of aqueous pore in a membrane, which facilitates the passage of ions. Of course such an aqueous pore (in principle just a hole) must meet all the thermodynamic requirements, meaning the hole must have a hydrophilic inner face to enable contact with the hydrophilic aqueous solution and a hydrophobic outer face for contact with the hydrophobic membrane core. Proteins can be arranged in three-dimensional structures within the membrane to create such holes. A diagram of a possible simple ionic-pore structure is given in Figure 2.9.

Comparable requirements to those stated to be necessary for transport in general are needed in the discussion of ionic-pores, too. A simple permanent open hole in a membrane would not be very useful for a living cell. The pore must have certain properties:

1. it must be selective
2. mechanisms must exist to control its activity
3. it must be able to open and close (gating)
4. pores can be precisely located in the membrane by cytosolic proteins.

A more sophisticated model of a membrane pore with all the required features is shown in Figure

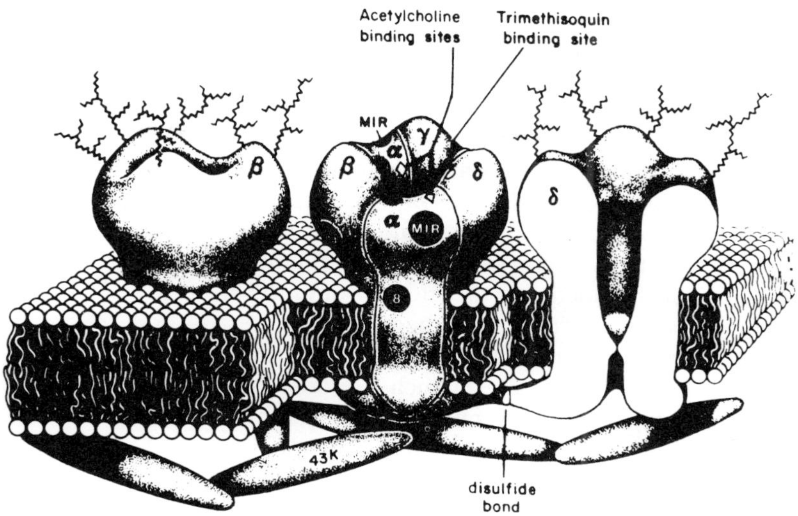


**Figure 2.10** An ion channel formed by a protein in a biological membrane is shown. This model includes the selectivity filter, the tunnel, the gate, binding sites for drugs and the envelope of the channel. MIR is the main immunogenic region and 8 marks a binding site for a monoclonal antibody. This example is the nicotinic acetylcholine receptor (nAChR) chosen here as it is the best described membrane channel. Receptor subunits α, β, γ, δ are indicated, as is the 43 kD protein associated with synaptic nAChRs. (Reproduced and modified from Anholt *et al.*, 1984 with permission.)

2.10. This is the proposed three dimensional structure of the muscle nicotinic acetylcholine receptor (nAChR), including much of the information currently available on the structure of this ionic channel (Kistler *et al.*, 1982). This model is partially speculative due to the fact that not very much is known about the three-dimensional structure of membrane proteins on the nanometre scale, nevertheless it may be useful to sketch the functional parts a protein must have in order to form an ion channel: hydrophilic core, selectivity filter, gate, binding sites for drugs and endogenous signals, and the envelope.

Such a membrane channel (pore) may be, at its simplest, considered to be a tunnel filled with aqueous solution; ions can move through it, following chemical or electrical gradients. Again the general laws of thermodynamics and electrodynamics hold. Very often, however, the pore does not behave like this, due to the fact that

```
1
Ac-AIB
   \ 2   3   4   5   6   7   8   9   10  11  12  13
    PRO-AIB-ALA-AIB-ALA-GLN-AIB-VAL-AIB-GLY-LEU-AIB
                   (AIB)                             \
                                                      PRO 14
                                      (GLN)          /
                          Pheol-GLN-GLU-AIB-AIB-VAL
                                19  18  17  16  15
```

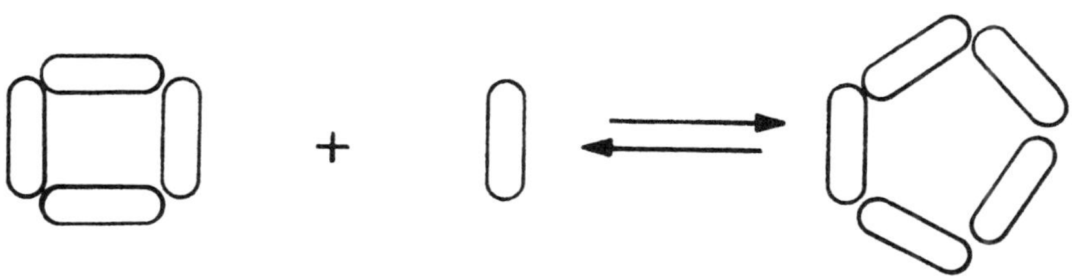

**Figure 2.11** Chemical structure of alamethicin (Boheim *et al.*, 1983) together with a diagram of the pore-forming mechanism of this polypeptide as suggested by Boheim (1974). The alamethicin pore is a homo-oligomeric aggregate which switches beween different states by association and dissociation of monomers to the active pore. Each monomer in the pore forms an α-helix crossing the entire membrane.

ions may interact with the tunnel lining, especially when the core of the channel is narrow. Theories have been developed to describe this finding, modelling pores as a series of energy barriers that the ion must cross. Such theories have the advantage that they are able to describe the selectivity of pores (Hille, 1992). Besides the permeation properties of an ionic pore, the gating properties of such pores (channels) must be described by adequate theories. Today, the gating of pores is widely described in terms of the different states a pore can adopt (Colquhoun and Hawkes, 1981, 1982). In such a model, a pore can jump between a number of different states in a stochastic manner, each of which may be conducting or non-conducting. The transition between the different states can be independent or regulated by external parameters such as membrane potential, the presence of so-called agonists [acetylcholine, γ-aminobutyric acid (GABA), etc.] or concentrations of ions (e.g. $Ca^{2+}$) or protons in the aqueous solutions.

In the following section we will present two examples of such pores in more detail as models for the general properties and structures of ion channels (ionic pores). Some additional data concerning pores and ion channels of biological relevance are given in Appendix 3.

The polypeptide antibiotic *alamethicin* forms voltage-dependent pores in artificial as well as biological membranes (Boheim, 1974). Alamethicin is a polypeptide antibiotic of 20 amino acids; the sequence is given in Figure 2.11 (upper). It is of amphipathic structure, and is thus both water- and lipid-soluble. Added to the aqueous solution bathing a membrane it diffuses into the membrane and there forms voltage-dependent pores. Information available to date indicates that these pores are formed by aggregation of some alamethicin monomers to build a homo-oligomer. The principal model of alamethicin pore formation is given in Figure 2.11 (lower). Each monomer in the membrane forms an α-helix with a hydrophobic surface and a hydrophilic surface which just spans the membrane. In the aggregate the hydrophilic surfaces are oriented inwards providing the hydrophobic tunnel of the pore. The gating of the pore is created by the change of the number of monomers which are integrated in the pore. As each α-helix has a dipole moment it will react

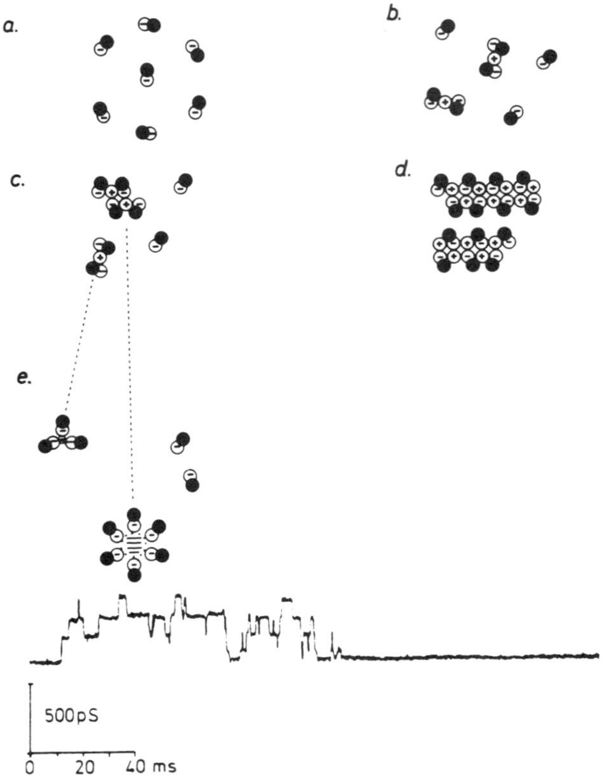


**Figure 2.12** Upper: Flip-flop model of alamethicin gating. Upon association to form the aggregate of the active pore, the entire α-helix of the alamethicin molecule is turned around in the membrane due to the applied electrical field. The model starts with the α-helical part of the alamethicin molecule being inserted into the membrane. Such monomers are shown in (a), exposing the negative pole of the molecule (*cis*-side), the non-helical part of the *cis*-side of the molecule is represented by a black dot. As long as this orientation is maintained only monomers can exist because of electrostatic interaction. However, when a molecule flips into an antiparallel orientation the situation changes, as antiparallel helices can now aggregate (b). When enough antiparallel molecules are present, larger aggregates may be formed ((c) and (d)). The final situation is shown in (e) when a proper membrane potential is applied to the aggregates in (c). The α-helices are forced by the electric field into a parallel configuration again, and because parallel α-helices repel each other, a situation as given in (e) may be reached. The hole between the alamethicin molecules may then be filled by water and an aqueous pore is formed. The strongly voltage-dependent gating behaviour of the pore formation of alamethicin is explained by this. Lower: Typical alamethicin-induced current fluctuation in a planar lipid bilayer in physiological salt solutions. The complete trace is of the activity of *one* alamethicin pore aggregate which switches between different conducting states. (Reproduced and modified from Boheim *et al.*, 1983 with permission.)

**Figure 2.13** Primary sequence of all subunits (α, β, γ, δ) of the nicotinic acetylcholine receptor molecule from the electrical organ of *Torpedo*. (Reproduced from Anholt *et al.*, 1984, with permission.)

with an electric field across the membrane, and the present model envisages that the whole α-helix is turned around in the membrane when being integrated into the active pore (Boheim *et al.*, 1983, 1984; Woolley and Wallace, 1992). A model of this behaviour is given in Figure 2.12 (upper), together with a recording of current fluctuations recorded from a single alamethicin pore (Figure 2.12 lower). Of course, the gating of other ionic channels may be quite different from that exhibited by alamethicin, but as a model for studying structure-function relations it is convenient and easy to handle when used together with planar lipid bilayers. Therefore the alamethicin pore in planar lipid bilayers will be described in some detail later in this book.

Most of the biological membrane channels are constructed quite differently from the alamethicin model. The native ion channel best understood to date is the nAChR which we already have shown in Figure 2.10. We will point out some basic properties of ion channels in general using the nAChR as a model. Due to its importance and the amount of work which has gone into its investigation, the reconstitution of this protein into planar lipid bilayers will be described later in this book in some detail (Section 8.3).

The structure of the nAChR has been explored using molecular biology methods. In Figures 2.13 and 2.14 the primary sequence of all the protein subunits is given, together with a simplified, hypothetical model of the three-dimensional structure of this protein in the membranes (see, for example, Noda *et al.*, 1983; Montal *et al.*,

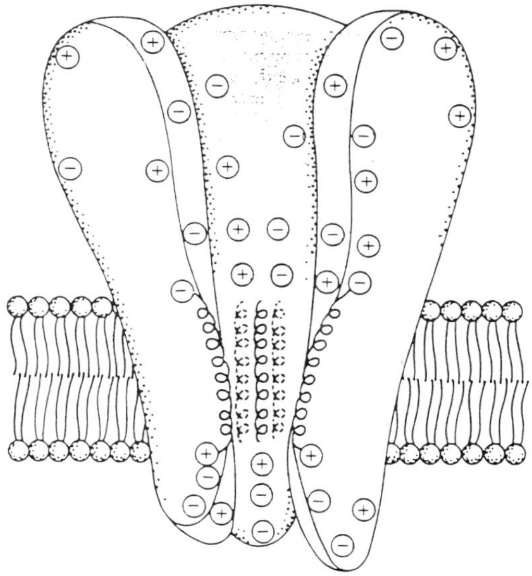

**Figure 2.14** Simplified three dimensional structure of the nicotinic acetylcholine receptor. The real three dimensional structure of this molecule *in situ* is not known, but from the data available a model such as that shown in Figure 2.10 has already been postulated. Here a simplified model is given. It is composed of all four subunits, with a stoichiometry $\alpha_2\ \beta\ \gamma\ \delta$, forming a pentamer.

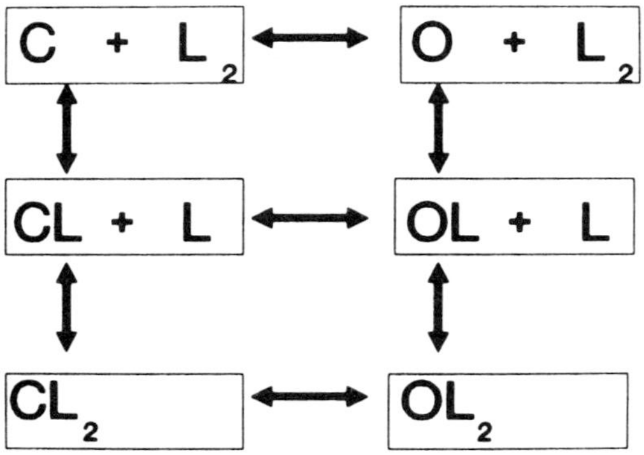


**Figure 2.15** Gating model of the acetylcholine receptor channel as deduced from the fact that two ligand molecules can bind to the two α-subunits of the AChR structure, thus opening the channel. O defines the open state of the channel, C defines the closed state of the channel and L shows whether ligands are bound in each of the states.

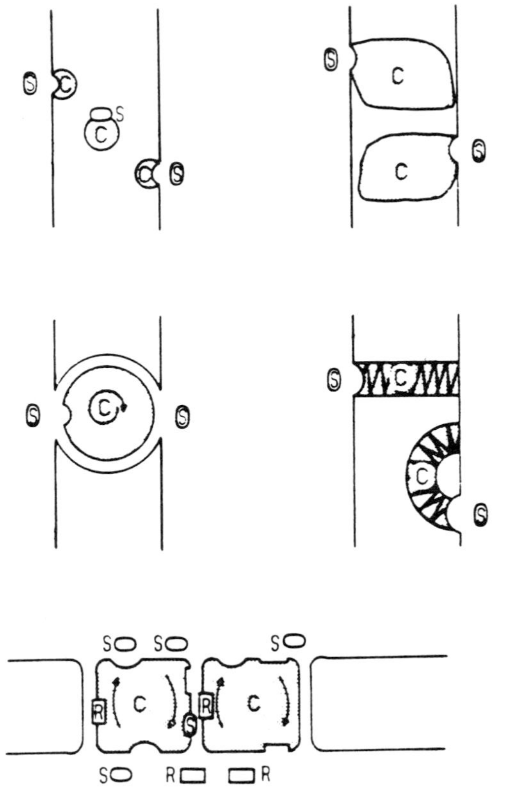

**Figure 2.16** Hypothetical models of ion carriers. Here some ideas are shown as to how ions can be transported by a carrier molecule across a membrane. Not all of these models can be realized in nature, but they demonstrate the variability of structures possible for one and the same task. C is the carrier molecule, R and S are substrates. (Reproduced from Höfer, 1977, with permission.)

1986; Hanke and Breer, 1989; Hucho, 1986). The nAChR is made from four different subunits (α, β, γ, δ), one of which (α) is represented twice in the entire aggregate. Each subunit forms a defined number of α-helices which cross the membrane. The activity of the channel is controlled by the binding of acetylcholine to the protein. Based on the finding that two acetylcholine molecules bind to the two α-subunits of the nAChR, a gating model has been proposed as shown in Figure 2.15, including open and closed states with up to two ligand molecules bound.

Although a lot is known about the structure and the function of the nAChR, many questions are still open as to how this channel really functions. Nevertheless, many of its basic properties are shared by other biological ion channels. Therefore it may serve as a general model for some basic properties of ion channel construction.

### *2.3.2.2 Ion carriers*

To enable an ion to cross a membrane it can be bound to an integral membrane protein which then moves the ion (integrated in a membrane-soluble complex) across the membrane by a conformational change or by movement of the whole complex. This can be done by different mechanisms and we present a collection of possible models in Figure 2.16. All these transport modalities, called carriers, share some common properties. Usually they are highly specific for one certain ion species (or molecule) due to their structure. The ion flux which is facilitated by these carriers follows electrical and chemical gradients across the membrane. Mechanistically such a system can be described by models, as in the very simple example shown in Figure 2.17. Here we have separated the association and dissociation of the ion and the carrier molecule from the transport across the membrane. The simple model in Figure 2.17 includes all the steps of ion transport by carriers. We will not go into more theoretical details here,

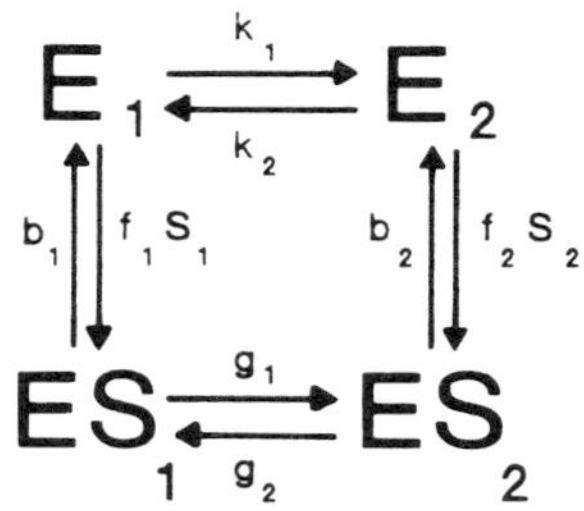


**Figure 2.17** Theoretical model to describe the ion transport across a bilayer by a carrier molecule. The model includes the association of the ion with the carrier, the transport of the complex across the membrane, the dissociation of the ion and the transport of the free carrier across the membrane. Each of the steps is reversible and it is important for this model that both the ion-carrier complex and free carrier molecule must be soluble in the lipid membrane and able to cross it. E is the carrier molecule, S the substrate, b and fS define binding and dissociation constants and k and g define transition rates.

the interested reader is referred to other textbooks and the original literature (Benz and Läuger, 1974; Crane, 1977; Höfer, 1977; Stein, 1990). A very well-described ion carrier is valinomycin, a $K^+$-selective ion carrier. The structure of this carrier is given in Figure 2.18 together with a three-dimensional model of the carrier–ion complex.

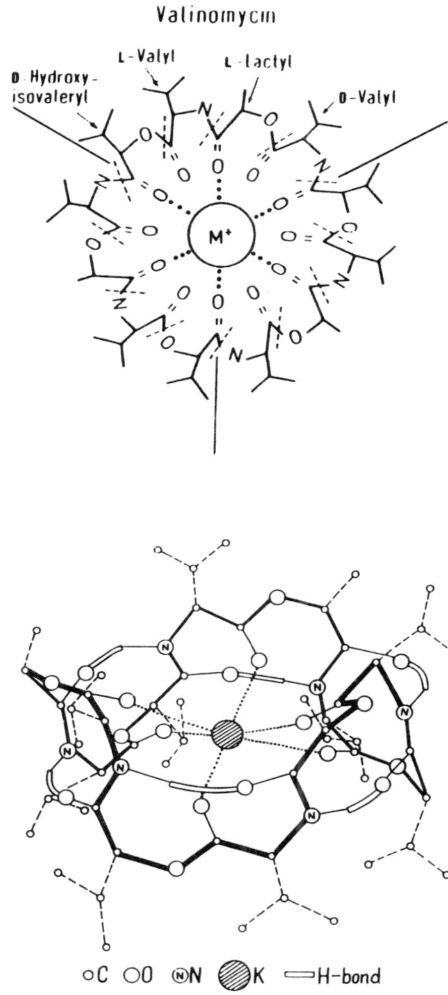


**Figure 2.18** Chemical structure of the $K^+$-selective carrier valinomycin (upper) together with a diagram of its three dimensional structure in the liganded and non-liganded state (lower).

### *2.3.2.3 Active transport*

In living cells material must often be transported against existing electrochemical gradients. The underlying process is often called active transport, although this term should be used in a more specialized way, to mean transport against a concentration gradient only. In any case, to achieve active transport, energy is necessary, which must be delivered in biological systems from the metabolism of the cell. This can be done by directly energizing the transport mechanism using metabolic energy, or by using the energy of the transmembrane gradient of another species of ion (or molecules) in the aqueous solutions. The direct use of metabolic energy is often called primary active transport, the use of energy from existing gradients is often discussed together with co-transport and counter-transport (Stein, 1988; Blostein, 1989; Fröhlich, 1989).

It is obvious that a simple membrane pore alone cannot fulfil the requirements of active transport as it only allows passive transport following a gradient. All structures now known to achieve active transport are similar to carriers, but additionally include an energy-consuming step. In primary active transport, this step very often involves the direct use of ATP (Skou *et al.*, 1988). More generally, energy can alternatively be delivered from an electric field across the membrane to support active transport, or from a gradient in concentration of a compound other than the transported one. All these types of mechanism have been found to exist in living cells. We will not go into details here but list a collection of some systems described in the literature:

1. $Na^+/K^+$-ATPase (Stein, 1988)
2. $H^+$-ATP-ase/synthase (Deleage *et al.*, 1983)
3. $Ca^{2+}/Na^+$ exchanger (Reeves, 1985)
4. bacteriorhodopsin and related molecules (Stoeckenius and Bogomolni, 1982; Gräeber, 1987).

There are many more. The interested reader is referred to the literature (see Introduction and Appendix 5). Some data on well-described carriers and pumps are given in Appendix 3.

## REFERENCES


Adam, G., Läuger, P. and Stark, G. (1977) *Physikalische Chemie und Biophysik*. Springer-Verlag, Berlin.

Benz, R. and Läuger, P. (1974) Kinetic analysis of carrier-mediated ion transport by the charge-pulse technique. *J. Membr. Biol.* **27**, 171–191.

Benz, R. and Zimmermann, U. (1980) Pulse-length dependence of the electrical breakdown in lipid bilayer membranes. *Biochim. Biophys. Acta* **597**, 637–642.

Benz, R., Becker, F. and Zimmermann U. (1979) Reversible electrical breakdown of lipid bilayer membranes: A charge–pulse relaxation study. *J. Membr. Biol.* **48**, 181–204.

Blostein, R. (1989) Ion pumps. *Curr. Opin. Cell Biol.* **1**, 746–752.

Boheim, G. (1974) Statistical analysis of Alamethicin in black lipid membranes. *J. Membr. Biol.* **19**, 277–303.

Boheim, G., Hanke, W. and Jung, G. (1983) Alamethicin pore formation: Voltage-dependent flip-flop of α-helix dipoles. *Biophys. Struct. Mech.* **9**, 181–191.

Boheim, G., Hanke, W., Jung, G. (1984) The Alamethicin pore is formed by a voltage-gated flip-flop of α-helix dipoles. In: *Chemistry of peptides and proteins*, Vol. III, W. Voelter, E. Bayer, Y.A. Ovchinicov and E. Wünsch (eds), Walter de Gryter, Berlin, pp. 281–289.

Colquhoun, D. and Hawkes, A.G. (1981) On the stochastic properties of single ion channels. *Proc. R. Soc. Lond. B* **211**, 205–235.

Colquhoun, D. and Hawkes, A.G. (1982) On the stochastic properties of bursts of single ion channel openings and of clusters of bursts. *Phil. Trans. R. Soc. Lond. B* **300**, 1–59.

Crane, R.K. (1977) The gradient hypothesis and other models of carrier-mediated active transport. *Rev. Physiol. Biochem. Pharmacol.* **78**, 99–159.

Deleage, G., Penin, F., Godinot, C. and Gautheron, D.C. (1983) Correlation beween ATP hydrolysis, ATP synthesis, generation and utilisation of delta pH in mitochondrial ATPase–ATP synthase. *Biochim. Biophys. Acta* **725**, 464–471.

Einstein, A. (1956) *Investigations on the theory of the Brownian movement*. Dover Publications, New York.

Findlay, J.B.C. and Evans, W.H. (1987) *Biological membranes: A practical approach*. IRL Press, Oxford.

Froehlich, O. (1989) Antiporters. *Curr. Opin. Cell. Biol.* **1**, 729–734.

Graeber, P. (1987) Primary charge separation and energy transduction in photosynthesis. In *Bioelectrochemistry* II. Milazzo and M. Blank (eds). Elsevier North Holland, Amsterdam, pp. 379–429.

Hanke, W. and Breer, H. (1989) Reconstitution of acetylcholine receptors into planar lipid bilayers. In *Subcellular biochemistry*, Vol. 14. J.R. Harris and A.H. Etemadi (eds). Plenum Press, New York, pp. 339–362.

Hille, B. (1975) Ion selectivity of $Na^+$ and $K^+$-channels of nerve membrane. In G. Eisenmann, (ed.)., *Membranes*, Vol. 3. Dekker, New York.

Hille, B. (1992) *Ionic channels of excitable membranes*. Sinauer, Sundreland, Massachusetts.

Höfer, M. (1977) *Transport durch biologische Membranen*. Verlag Chemie, Weinheim.

Hucho, F. (1986) The nicotinic acetylcholine receptor and its ionic channel. *Eur. J. Biochem.* **158**, 221–326.

Kistler, J. Stroud, R.M., Klymkowsky, M.W., Lalancette, R.A. and Fairclough, R.H. (1982) Structure and function of an acetylcholine receptor. *Biophys. J.* **37**, 371–383.

Lindau, M. and Gomperts, B.D. (1991) Techniques and concepts of exocytosis: focus on mast cells. *Biochem. Biophys. Acta* **1071**, 429–471.

Montal, M., Anholt,R. and LaBarca, P. (1986) The reconstituted acetylcholine receptor. In: C. Miller, (ed.). *Ion channel reconstitution*. Plenum Press, New York.

Noda, M. Takahashi, H., Tanabe, T., Toyosato, M., Kikotani, S., Furutani, Y., Hirose, T., Takamisha, H., Ianayama, S., Miyata, T. and Numa, S. (1983) Structural homology of *Torpedo californica* acetylcholine receptor subunits. *Nature* **302**, 528–532.

Parsegian, A. (1969) Energy of an ion crossing a low dielectrical membrane: Solutions to four relevant electrostatic problems. *Nature* **221**, 844–846.

Potter, H., Weir, L. and Leder, P. (1984) Enhancer dependent expression of human K immunoglobulin genes introduced into mouse pre-B lymphocytes by electroporation. *Proc. Natl. Acad. Sci. USA* **81**, 7161–7165.

Reeves, J.P. (1985) The sarcolemmal sodium-calcium exchange system. *Current topics in membranes and transport*, Vol. 25, 77–127.

Skou, J.C. et al. (eds) (1988) *The $Na^+$, $K^+$-pump. Parts A and B*. Alan R. Liss, New York.

Starzak, M.E. (1984) *The physical chemistry of membranes*. Academic Press, New York.

Stein, W.D. (1988) *Ion pumps: Structure, function and regulation*. Alan R. Liss, New York.

Stein, W.D. (1990) *Channels, carriers and pumps: An introduction to membrane transport*. Academic Press, New York.

Stoeckenius, W. and Bogomolni, R.A. (1982) Bacteriorhodopsin and related pigments of *Halobacteria*. *Annu. Rev. Biochem.* **51**, 587–616.

Trowbridge, I.S. (1991) Endocytosis and signals for internalization. *Curr. Opin. Cell Biol.* **3**, 634–641.

Woolley, G.A. and Wallace, B.A. (1992) Model ion channles: Gramicidin and alamethicin. *J. Membr. Biol.* **129**, 109–136.

CHAPTER THREE

# Technical details of bilayer experiments

Independent of any theoretical reflections, the basis of successful bilayer experiments should be reliable technical equipment. In principle not a great deal is needed in terms of technical support to perform good bilayer experiments. We have already presented the fundamental ideas of a classical bilayer set-up in Figure 1.4; in this chapter we want to start with a more expanded view of a practical bilayer set-up, containing all the parts necessary for a beginner to embark on this technique. In Figure 3.1 a diagram of a complete bilayer system is shown containing a variety of components, mainly mechanical, electrical and optical items. In the following section we will discuss this equipment in detail and provide information on its use in bilayer work (Ti Tien, 1974; Latorre, 1986, Miller, 1986).

## 3.1 MECHANICS

### 3.1.1 Vibration isolation

A bilayer is a very fragile mechanical system, a flat area of some 100 $\mu m^2$ up to 1 $mm^2$, which is about 5 nm thick. Thus it is extremely sensitive to mechanical disturbances and vibrations. A good bilayer set-up therefore must be placed on a vibration isolation table. A straightforward solution is to buy such a table from commercial sources, and we have listed the addresses of some sources at the end of the book (Appendix 4). Usually the price of such tables (which are widely used in optical experiments, for example) is quite high. A cheaper solution is to build the table from simple components. The first part one needs is a heavy stone plate (or metal plate), the dimensions of which should be about 60 × 40 × 10 cm; such a plate will have a weight of about 50–100 kg. This plate can be placed on four tennis balls, for example on top of a normal laboratory

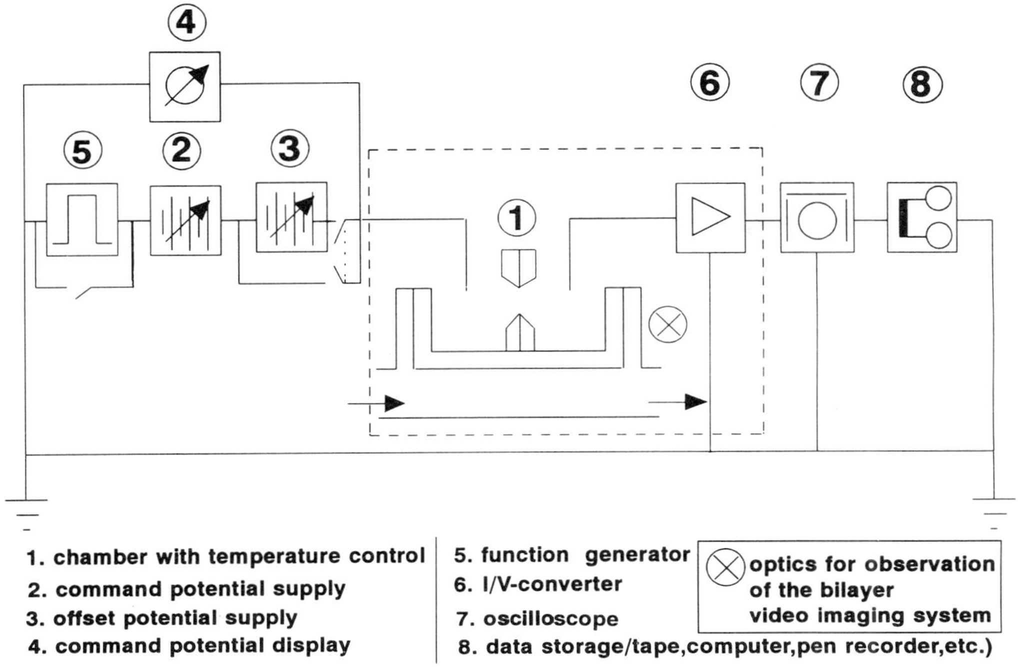


**Figure 3.1** A complete bilayer set-up with all its main components: mechanical, optical and eletronic items.

table, and this alone will give a quite good vibration isolation. Alternatively, one can use a partially inflated inner tube from a vehicle to support the plate. In Figure 3.2 a vibration isolation table used with good results by the authors is illustrated. Here pneumatic shock absorbers were used instead of tennis balls.

### 3.1.2 Faraday cage

Due to the fact that in bilayer experiments very low currents have to be measured across the bilayer, highly sensitive amplifiers must be employed. This makes it necessary to shield part of the set-up by a Faraday cage. In bilayer set-ups the chambers and the headstage of the current to voltage converter (I/V-converter) are usually placed within a Faraday cage. Additionally, the devices used for temperature control of the bilayer chamber and all equipment that is in direct contact with the aqueous solutions must be shielded. It is best to build a metal box around the components mentioned above. The dimensions of this box are of course dependent on the size of the particular apparatus. We used a standard size cage of 20 × 30 × 15 cm. The box must have a door at one side to give access to the interior.

The box can be made of solid plates. These plates must have a proper thickness because when using very thin plates the walls of the box themselves become sensitive to acoustic vibrations. Plates 3–5 mm thick give good results when steel or aluminium is used. Alternatively, the walls of the box can be made from metal nets which must be mounted securely. Two examples of such Faraday cages are shown in Figure 3.3.

### 3.1.3 Experimental chambers

Usually the main part of a bilayer set-up is a chamber divided in two parts, which contains the aqueous solution bathing both sides of the artificial membrane. A variety of such chambers have been used in different laboratories, and we want to present a few models here. When constructing such a chamber several features have to be considered:

1. the type of bilayer for which the chamber will be used (see also Chapter 5)

**Figure 3.2** A vibration isolation table (home-built) used by the authors in a bilayer set-up. Pneumatic shock absorbers are used for vibration isolation of a heavy stone plate. This should be taken as only one example; numerous different designs have been shown in the literature and are commercially available. FC, Faraday cage; SA, shock absorber; SP, stone plate.

**Figure 3.3** Two different types of Faraday cage used for bilayer set-ups. One set-up consists of a closed metal cage (aluminium), the other is shielded by a thin metal net.

2. the size and the geometry of the chamber
3. the material for constructing the chamber
4. the construction of the partition to divide the chamber into two compartments
5. the procedure to make the membrane carrying aperture (this is described in more detail later).

With regard to point 1, one first has to decide what type of bilayer will be needed. In Chapter 5 we will describe the different types of bilayer techniques in detail; in the present chapter we will simply state the type of bilayer the chambers are suitable for, using the terminology of chapter 5 to define the different bilayer types.

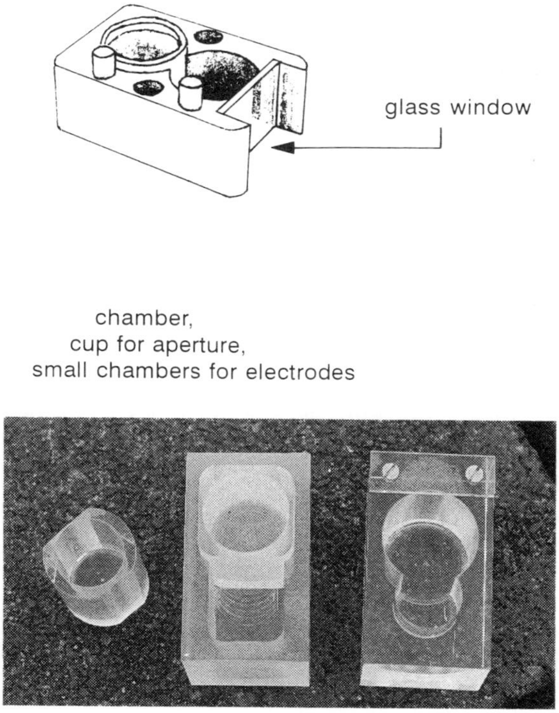


**Figure 3.4** Chamber for bilayer experiments, mainly to be used for painted bilayers. The diagram at the top shows a small plastic cup fitted into a block also made of plastic. In this block two holes are made, one for the cup and the other one to serve as an opposite chamber (for easier handling, usually used as the front chamber). The small cup has the membrane carrying-aperture on one side, placed to the front. The lower part of this figure shows a photograph of chambers used by the authors with good success. In one of these chambers the small cup additionally is prevented from turning around by a clamp.

For *painted bilayers*, the chambers are best made as shown in Figure 3.4. A diagram of such a chamber is given in this figure. It consists of two parts, the basic block, in which two holes are drilled and a cup which fits into the rear hole. The basic block can be made from almost any plastic material. When Teflon is used it has the advantage of easy cleaning. In case optical observation of the bilayer is required, the front part of the basic block must be supplied with a quartz window (not drawn). In the wall of the cup a small hole is made to hold the bilayer. For ease of use, the round cup should be fixed in the block by a proper clamp. The diameter of the membrane-carrying aperture may range from 0.1 to 2.0 mm. When the diameter of this aperture is in the millimetre range the cup is best made from Teflon; at smaller apertures, the cup should be made from polycarbonate or comparable material as painted bilayers of small size do not spread very well and become black films on Teflon. Figure 3.4 also shows photographs of chambers of this type. Such chambers have been used by the authors with good results.

*Folded bilayers* are usually made at smaller size (diameter $<0.3$ mm). Additionally, this type of bilayer requires thinner partitions than those used with painted bilayers. In Figure 3.5 we present two types of chambers fitting the needs of folded bilayers. The type presented on the left is made from two chambers (made from Teflon) which have bigger holes (diameter some 1 cm) in the front. Between these two chambers a Teflon film (some 5–20 μm thick) with the

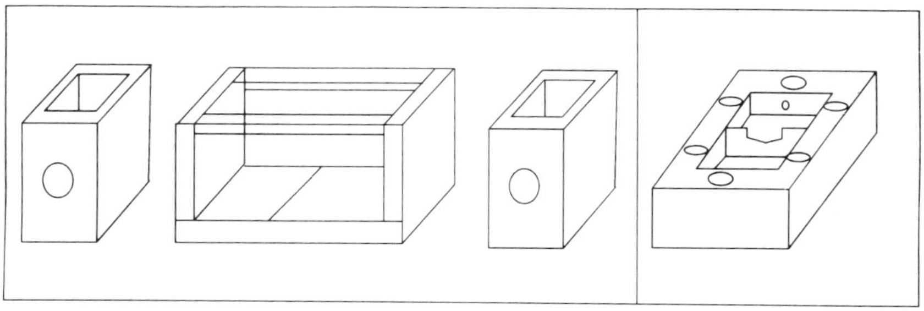

**Figure 3.5** Two different types of bilayer chambers to be used mainly with folded bilayers. Left: a system composed of two separate parts which must be clamped together by an appropriate device. A Teflon film with a small aperture is fixed between the two parts of the complete chamber. Right: a system which uses one chamber divided by a wall. Into this wall a V-shaped hole is cut, on which a Teflon film can be glued with silicon grease.

membrane-carrying aperture (diameter about 0.1–0.3 mm) is fixed by an appropriate proper metal clamp (the clamp is shown in Figure 3.6). To prevent leakage between the chambers, usually they are glued together with the Teflon film by silicon grease. The chamber type shown at the right side consists of a basic Teflon block which again is divided by a wall. A V-shaped hole is cut into this thin wall. Onto the V-shaped hole a Teflon film is glued by silicon grease, which contains the membrane carrying aperture. Both types of chamber work well, but one must handle them properly to prevent leakage. With the first type of chamber this is a little easier, as the components are fixed by a metal clamp (Figure 3.6).

In addition to the design of the chambers described so far, the basic block may contain two additional small holes to locate and fix the electrodes and agar bridges. Instead of these holes one can use separate small cups for the electrodes, which are placed close to the bilayer chamber itself. The use of agar bridges is of advantage over the use of simple Ag/AgCl electrodes, as differences in the chloride concentration in the two bathing solutions do not add a junction potential to the bilayer. Agar bridges can be made easily from glass capillaries (outer diameter 1.5 mm, inner diameter 1.2 mm) which are bent U-shaped at proper length (20 mm long and 5 mm wide). These capillaries are then filled with hot agar in 1 M KCl. After cooling, they are ready for use. The agar bridges should be rinsed carefully with water after every experiment and stored in 1 M KCl.

**Figure 3.6** Photograph of a disassembled chamber for use mainly with folded bilayers. The type sketched in the left part of Figure 3.5 is shown.

### 3.1.4 Partitions and apertures

A very important part of each bilayer experiment is the membrane carrying aperture (see also Chapter 5). It must be made with as small a boundary ripple as possible, to avoid disturbances of the bilayer.

When painted bilayers are used, the hole is usually drilled into the partition (wall of the cup) dividing the two parts of the bilayer chamber. A small drill of appropriate diameter can be used for this purpose. Remaining boundary ripple is compensated for at least partially by the amount of organic solvent (decane) in the system. An improved shape of an aperture to be used for painted bilayers is given in Figure 3.7; this shape gives better stability of the bilayer during an experiment. Nevertheless, a simple hole will do nearly as well.

When folded bilayers are used, the membrane-carrying aperture is usually made into a Teflon film of a thickness of 5–25 μm. This can only be achieved by appropriate tools. One way to make such a tool, is to cut an injection needle of the

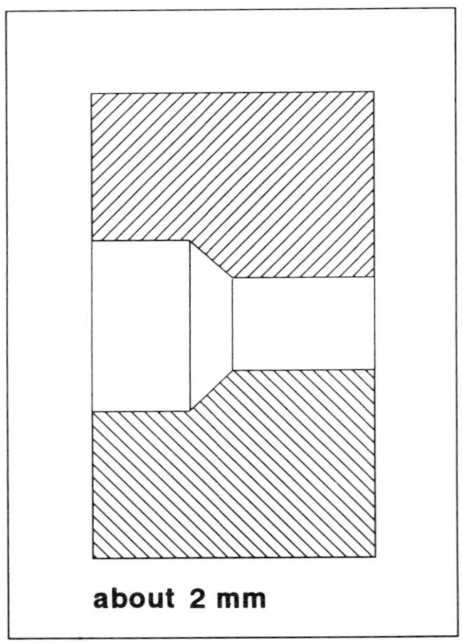


**Figure 3.7** Simple small holes in plastic partition are used usually as apertures for bilayer experiments (painted bilayers). In this figure we show the shape of an aperture which will better support the long-term stability of painted bilayers.

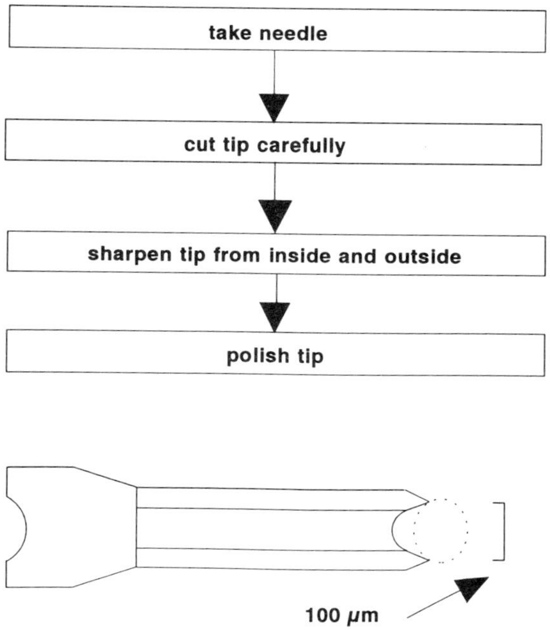


**Figure 3.8** A tool that can be used to punch small holes (in the range of 100 μm diameter) in a Teflon film. A flow chart shows how to make this tool from a needle of a syringe.

diameter needed at the tip and to sharpen the edge as shown in Figure 3.8. With such a needle one can punch reproducible holes of a diameter from 0.1–0.3 mm into Teflon film. Another way is to burn the hole into the Teflon film with a high-voltage spark. For this, a piece of film is mounted between the tips of two metal needles and a high-voltage pulse is applied to the needles. A set-up to do this is shown in Figure 3.9. The height of the potential (some thousand volts) and the length of the electrical pulse (some milliseconds) needed to perforate the film depends on the distance between the two needles, the thickness of the film and the quality of the film material. The values given in the legend of Figure 3.9 are typical values used by the authors. The electric pulse can be produced either with a high-voltage pulse generator (which will be quite expensive) or, for example, by an ignition coil (electronic or capacitor ignition system) from a car; this system is not as easy to handle but is much cheaper. In Figure 3.9 together with the photo of a set-up build by the authors an outline of such an electronic circuit is given.

When the Teflon film used is very thin, the mechanical handling of such a partition becomes more and more difficult. For this reason we have created a technique to stabilize the Teflon film by two supporting rings of thicker Teflon. Two rings of fibreglass coated Teflon which are 80 μm thick, are melted to a septum together with the piece of thin Teflon in between. The rings have an outer diameter of 20 mm and an inner diameter of 10 mm, the thin Teflon sheet has a diameter of 20 mm. Such a 'septum' made from the thin Teflon film and the two supporting rings has to be tight against leakage; a procedure to make such septa is given in Figure 3.10 in form of a flow chart. Finally, in Figure 3.11 a photo of a ready-to-use septum together with a microphotograph of the membrane-carrying aperture is given.

### 3.1.5 Cleaning procedures for chambers and partitions

Due to the fact that a bilayer is a very sensitive system for detecting membrane-soluble chemical compounds, the chambers and partitions used to make bilayers must be cleaned very carefully. The way this can be done best depends partly on the material the chambers are made from. There are a few general steps which should be always followed, however.

After every experiment chambers and partitions (if they are used more than once and not disposable) are best washed under running water to remove all salts and water-soluble compounds. Then they should be washed by ethanol or methanol; this removes most of the lipids and other hydrophobic material in the chambers. If the chambers are made from Teflon, instead of plain methanol a mixture of 2:1, v/v, methanol: chloroform can be used. After that, chambers must be carefully rinsed with quartz distilled water, to get rid of all remaining water-soluble compounds and of course the alcohol. If silicon grease has been used with the chambers, this must be removed first (with a paper tissue).

The next steps can only be performed with Teflon chambers. One can clean them with

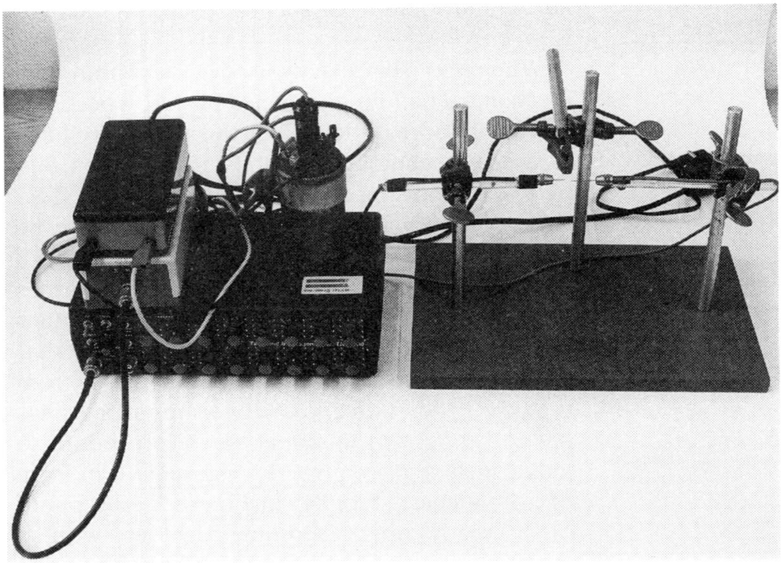

**Figure 3.9** A set-up used by the authors to burn small holes in a Teflon film. The piece of film is clamped between the two needles. Teflon film up to 25 μm thick can be used with this system. The voltage pulse is delivered by an ignition coil (some 10 000 V for 10 ms) from a car. To give a more defined spark with films thicker than 10 μm, it is useful to punch a very small hole into the film at the place where the final membrane carrying aperture is required, before applying a spark (this can be best done by a fine glass or metal needle with a tip diameter of about 10–100 μm). Below the photograph a simplified flow chart for the electronics is shown.

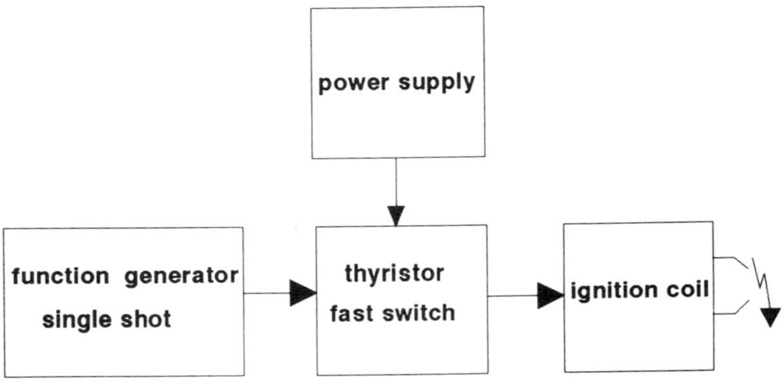


sulphur–chromic acid (overnight), which is discarded after the cleaning procedure. Then the chambers must be washed again for some hours in running water to remove the remaining acid. After that they can be treated by a 1 M solution of NaOH and again a wash under running water (instead of sulphur–chromic acid, for environmental reasons a mixture of HCl and $HNO_3$ can be used).

The chambers are finally boiled for about an hour in quartz-distilled water; it is better when this is done shortly before use of the chambers.

### 3.1.6 Perfusion system

One of the advantages of bilayer experiments is the easy access to both sides of the membrane. To really profit from this, an easy-to-handle system is needed to change the aqueous solution on both sides of the bilayer. As we have stated already, bilayers are mechanically very sensitive, therefore such a perfusion system must produce as little mechanical disturbance as possible. In addition, the perfusion system should be designed to guarantee that the level of the water surface is not changed during the perfusion procedure. We will here present some details of a hand-driven device which fulfils all the points required (Figure 3.12). In this device two identical syringes are used, which are coupled mechanically. By this, identical amounts of solution are added to the bath from one side and taken away from the other. The feed should be placed at the bottom the surface of the aqueous solution, for better solution exchange.

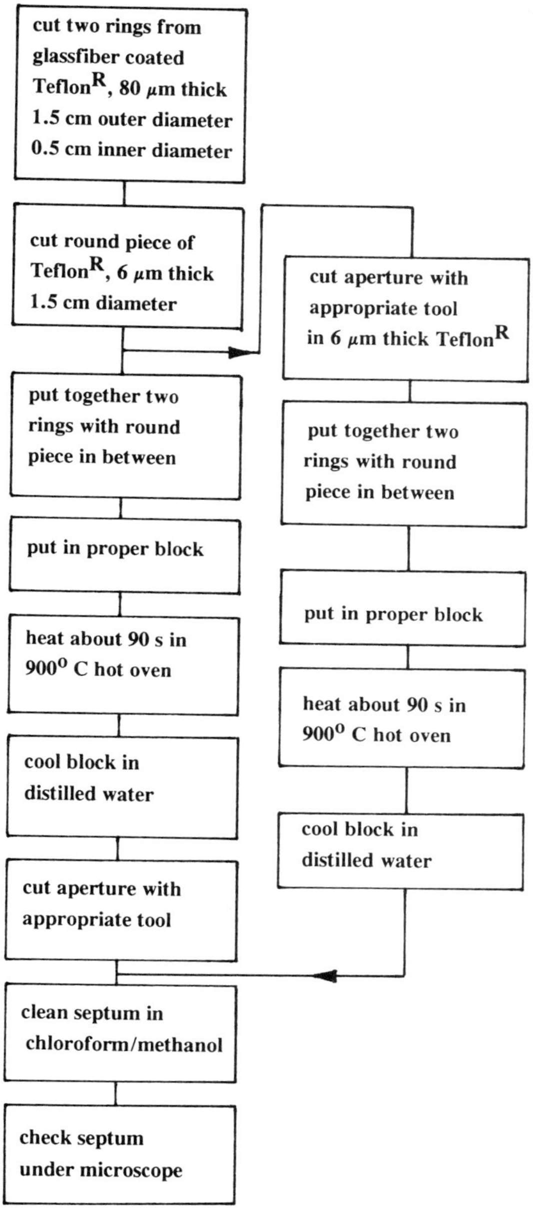


**Figure 3.10** Flow chart of a procedure to make septa for folded bilayers. The thin Teflon film which is used to carry the aperture can be 5–25 μm thick. The supporting rings are made from 80 μm thick glassfibre coated Teflon in the example shown. The diameter of the piece of thin Teflon film is 20 mm, the rings have an outer diameter of 20 mm and an inner diameter of 10 mm.

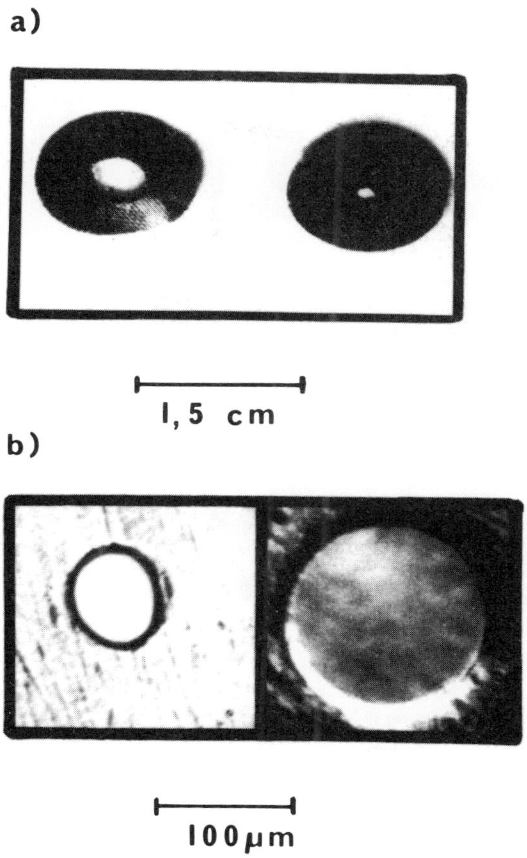


**Figure 3.11** Photo of a ready to use septum together with a microphotograph of the aperture in this septum. The dimensions of the septum are the same as those stated in Figure 3.10.

A good workshop will be able to build a motor-driven device for ease of operation, but the simple manual device will do as well. We have listed some sources of equipment for perfusion systems in Appendix 4.

### 3.1.7 Stirring devices

In some bilayer experiments it is necessary to stir the solutions bathing the bilayer. This can be best achieved by small magnetic stirrers of the type shown in Figure 3.13. In the example presented here, the magnetic stirrer is driven by a bigger magnet placed on a motor below the bilayer chamber. Of course the stirrers must be switched off, when performing electrical measure-

**Figure 3.12** Perfusion system for bilayer chambers. The perfusion system shown can be used to perfuse one bilayer chamber on a time-scale of some 10 s. Two syringes of identical volume are coupled mechanically and are used in a push–pull manner.

ments in order to prevent mechanical artifacts in the recordings.

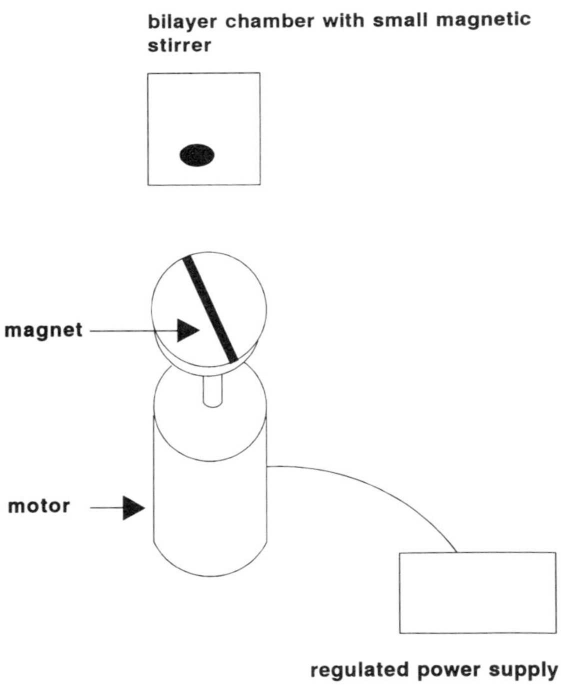


**Figure 3.13** A device for stirring aqueous solutions in bilayer chambers. A small magnetic stirrer (in the bathing solution) is driven by a magnet, placed on a motor below the bilayer chamber.

### 3.1.8 Temperature control and measurement

In bilayer experiments the temperature of the membrane can be controlled very easily. The chamber used for the bilayer experiment is best cooled or heated by a appropriate block into which the bilayer chamber fits. In early bilayer experiments, the chambers were fit into a hollow metal-block which is filled by water controlled by a thermostat. A photograph of such a system is given in Figure 3.14.

Instead of such a device, the bilayer chamber can be temperature controlled by a peltier-driven system, as well. Such systems are available from several sources (see Appendix 4). Custom-made miniature cooling and heating coils are commercially available, too. They can be immersed directy into bilayer chambers when these are of sufficient size. Another possibility is to insert a small coil made from a glass tube into the chamber and to perfuse this coil by temperature-controlled water.

To measure the temperature of a bilayer, the easiest way is to insert a small temperature probe into the aqueous solution bathing the bilayer.

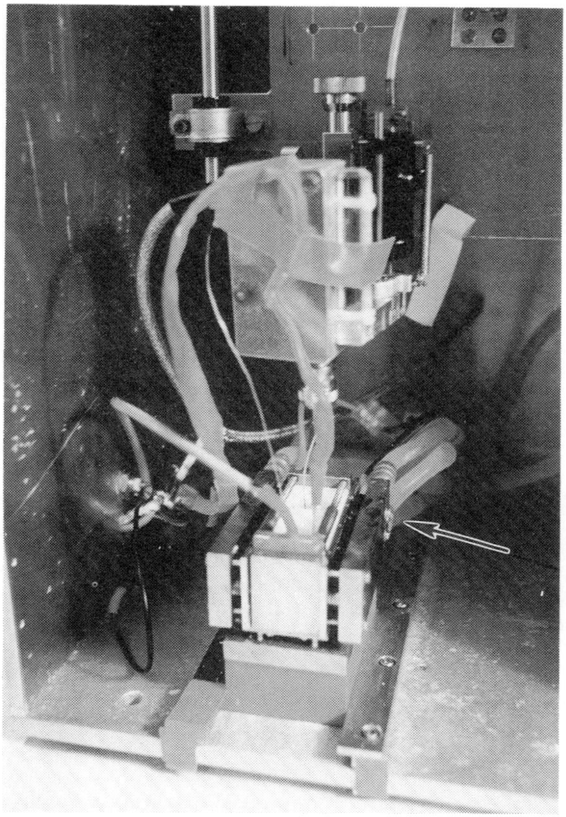

**Figure 3.14** Temperature control of a bilayer chamber. A bilayer chamber can be placed in a block which is perfused with water (or alcohol) by use of a thermostat to control the temperature of the bilayer. The arrow indicates the cooling–heating block.

The output of the temperature probe then can be used as the input signal for an electronic temperature regulation circuit. In Figure 3.15 a diagram shows how to build a complete temperature control system.

### 3.1.9 Measurement of pH

To measure the pH in the aqueous solutions of a bilayer experiment, any pH electrode which is sufficiently small will do. Alternatively ion-sensitive glass microelectrodes may be used, which are described in the literature in detail (Thomas, 1978; Corvington, 1979).

The same technical principles as for pH measurement can be used when the control of the concentration of another ion species (for example calcium) is required. The interested reader should refer to the original literature: for an introduction to the field see for example, Thomas (1978).

### 3.1.10 Equipment for dip-stick bilayers and patch-clamping large liposomes

Besides the classical methods to make bilayers (see Chapter 5) some new techniques have been developed as a result of the influence of the patch-clamp technique on the field of bilayer research (Hamill *et al.*, 1981). This makes it necessary to discuss at least the fundamentals of these applications. As they are identical to the patch-clamp technique in most of their experimental aspects we will, however, present the principles only and otherwise refer to the literature on patch-clamp experiments (e.g. Sakmann and Neher, 1983).

One advantage of the techniques summarized under the name dip-stick techniques is that any laboratory having a patch-clamp set-up available can use this set-up without any changes being necessary for bilayer experiments using glass microelectrodes.

#### *3.1.10.1 Dip-stick set-up*

When doing so-called dip-stick experiments, a pipette has to be moved close to the air–water interface of a bath solution (see Chapter 5). This can be done with or without optical control. The basic mechanical components needed are:

1. a coarse mechanical manipulator
2. a one axis microdrive
3. a pipette holder
4. a chamber for the bath solution.

The remaining parts and accessories are identical to those used in classical bilayer experiments. A diagram of a such a simplified dip-stick set-up is given in Figure 3.16.

#### *3.1.10.2 Patch-clamp set-up*

Patch-clamping large liposomes is fully identical to patch-clamp experiments on whole cells (Kirino

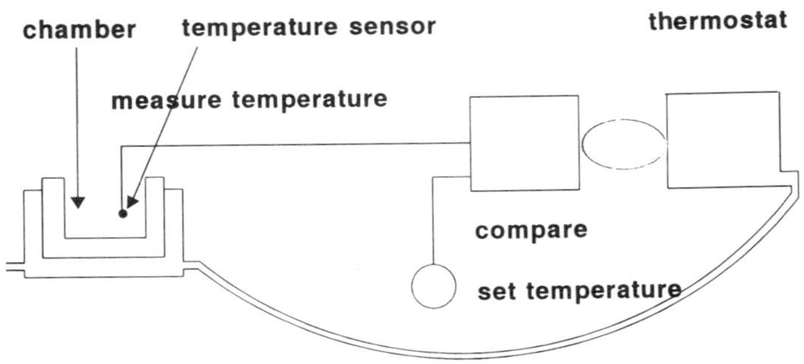


**Figure 3.15** A block diagram of a simple electronic temperature control system.

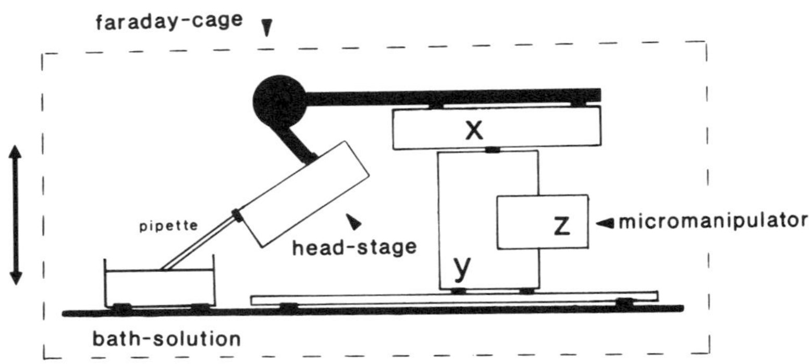


**Figure 3.16** Simplified design of a set-up to be used for dip-stick bilayers. Only those parts of the set-up which differ from systems used for other types of bilayer experiments are shown.

**Figure 3.17** Photo of a patch-clamp set-up built by the authors. This is just one example from an incredible number of different designs found in patch-clamp laboratories. Comparable devices are commercially available (see Appendix 4).

*et al.*, 1991). In Figure 3.17 for reasons of completeness a photograph of a patch-clamp set-up used by the authors is shown. Literature on patch-clamp equipment is readily available (Sakmann and Neher, 1983).

### *3.1.10.3 Pipettes*

We want to point out here that in bilayer experiments carried out with glass pipettes, the quality requirements of the pipette are usually lower than in the patch-clamp technique itself. Fire-polishing of the pipettes is usually not necessary. As most dip-stick experiments are done close to the surface or at the surface of the aqueous solution, coating of the pipette also is normally not required.

All the other requirements of patch-clamping large liposomes are identical to patch-clamp experiments on whole cells. Thus, for example, glass capillaries with filament are recommended for easier filling. An appropriate two-stage microelectrode puller is needed for the fabrication of the pipettes.

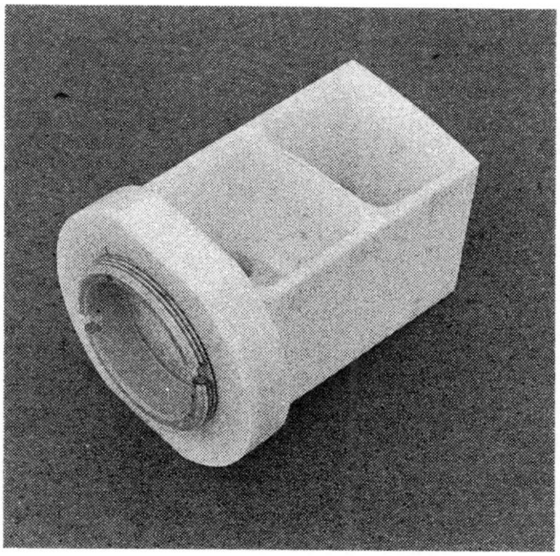

**Figure 3.18** A chamber is shown, which can be used for bilayer experiments under optical control.

## 3.2 OPTICS

In the earlier bilayer experiments, optical techniques were used exclusively for observing the thinning of coloured films from decane/lipid films to real black lipid bilayers. Independent of the beauty of these images, optical methods have now also become important for a variety of other investigations. Thus we will briefly describe some useful pieces of optical equipment for bilayer experiments in the following paragraphs (an introduction to optics methods in biology is given by Slayter, 1970).

### 3.2.1 Microscopes and telescopes

The direct observation of lipid bilayers, especially of decane/lipid bilayers is even today a useful tool:

1. it can be used to control the thinning of the film
2. it can be used to measure the real area of a lipid bilayer, which may be smaller than that of the aperture due to the annulus of lipid and organic solvent.

For direct observation, the bilayer chamber must have a window through which the membrane can be visualized. In front of this window a lamp must be installed for illumination of the bilayer and a microscope for observation of the membrane in the reflected light. The focal distance of the optics should be in the range of 10–20 cm for convenient handling, the magnification should be 10–20 times. To the optics a video camera can be mounted for even more comfortable observation of the membrane on a monitor. A diagram of the classical set-up to examine black-lipid-films is given in Chapter 5, where this type of bilayer is described in more detail. Here we only show a diagram of the components needed in a more sophisticated system (Figures 3.18 and 3.19) together with a chamber and a cage having windows for optical observation.

### 3.2.2 Fluorimetry

In the last decade fluorescent dyes have been developed for a wide range of applications (Haughland, 1992; Wilson, 1989). Some of them will bind to lipid membranes or proteins, others are soluble in aqueous solution. These dyes are sensitive for a variety of parameters, for example, membrane potential or calcium concentration. By the use of such dyes some properties of lipid bilayers can be studied quite easily.

In this chapter we present the basic technical requirements (according to the optics used) of these studies, without describing the methods for labelling lipid bilayers with dyes or for handling other dyes; the reader is referred to the original

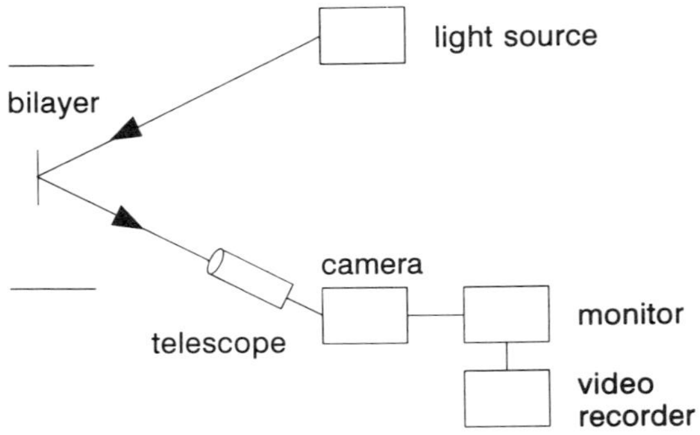


**Figure 3.19** In a more sophisticated set-up the optics for observing a bilayer might by used together with a video camera, a monitor and a computer based video imaging system.

literature for this (i.e. Niles and Cohen, 1991; Haughland, 1982).

In an experiment visualizing the change of fluorescence in the front chamber of a bilayer set-up or at the side of the bilayer itself, a bilayer chamber and a cage with a window as described in the previous chapter are needed. The bilayer is illuminated by UV light and fluorescence is observed via a barrier filter of proper cut-off wavelength. For observation of bilayers with fluorescent dyes a fluorescence microscope (mounted according to the needs of observing the bilayer) can be used together with an objective of long working distance (objectives with up to 20 mm working distance are available, see Appendix 4). In cases when the bilayer is illuminated through the objective, a UV objective is needed. As the level of fluorescence often is quite low, an intensified video camera (sensitive to low light levels) is recommended. It should be mounted to the optics using a zoom lens. The images are best stored on a video-tape recorder and later digitized and processed with a frame grabber and a good image processor (for an introduction to video-imaging techniques in microscopy, see Inoue, 1986; Russ, 1990).

Additionally, we must point out that video-imaging systems and microfluorimeters have been developed for use with patch-clamp experiments and other investigations. The same technology can be used with bilayers when the necessary optics are installed.

### 3.2.3 Tracing dyes

Sometimes, especially in membrane fusion experiments, another technique of optical observation is useful in bilayer experiments. The principles of such experiments are given in Figure 3.20. A liposome filled with some type of dye fuses with a bilayer and empties its contents to the *trans*-side of the membrane. There, the concentration of the dye can be traced by optical methods (Zimmerberg *et al.*, 1980). As an example the liposomes can be filled with $Ca^{2+}$ and the aqueous solution of the *trans*-chamber can contain Arsenazo III as a $Ca^{2+}$-indicator. A lot of other dyes are possible, for example fluorescent dyes can be used with the equipment as described in the previous chapter. Whatever the case, the optical system has to be designed according to the requirements of the dye used.

## 3.3 ELECTRONICS

The main aspect of bilayer experiments is the electronic parts needed to measure currents across bilayers, store the results and evaluate data. These parts are described in this section.

### 3.3.1 I/V-converter

One of the most important parts of any bilayer set-up is an I/V-converter to measure currents across the bilayer at a given membrane potential. The principles of these measurements are thus similar to classical voltage-clamp experiments (DuBovy, 1978; Fishman, 1982). The currents which have to be measured in bilayer experiments may range from some nanoamperes down to picoamperes when single ion channels are under investigation. This makes it necessary to use I/V-converters of sufficiently high sensitivity. With commercially available operational amplifiers it is not difficult to build such an amplifier by oneself.

Of course one can buy a ready-to-use amplifier. The amplifiers which are sold for bilayer experiments very often have some nice

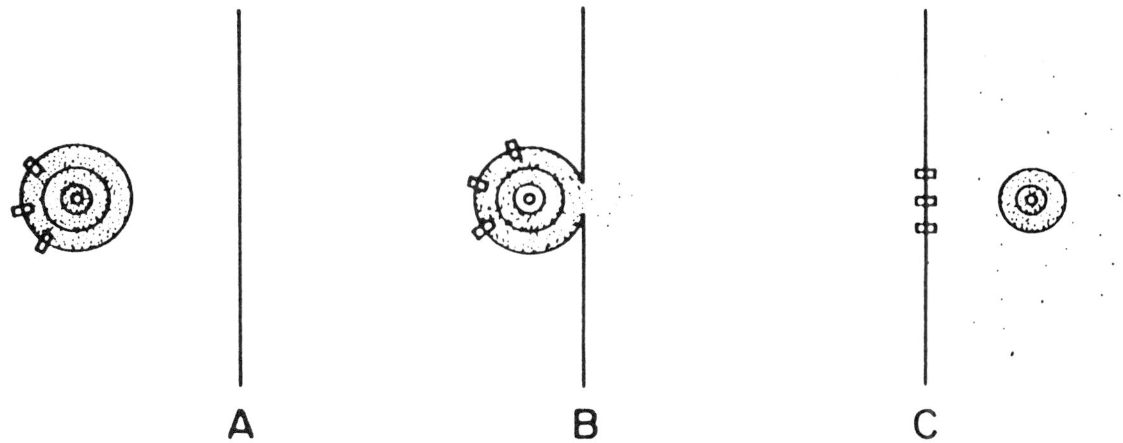


**Figure 3.20** Fusion of liposomes with a bilayer can be optically traced as shown in this Figure. The vesicles are added to the *cis*-side of the aqueous solution bathing the bilayer. They are preloaded with calcium for example. The *trans*-side of the aqueous solution bathing the bilayer is calcium-free. Upon fusion events, the contents of each fusing liposome is emptied across the bilayer to the *trans*-side. The increase of calcium in the aqueous solution of the *trans*-side can be visualized by proper dyes. Together with calcium, Arsenazo III (or Antipyrelazo III), for example, can be used. The shift in the absorbance of light by the Arsenazo III solution can be measured optically as sketched in the figure.

additional features, as there are built-in command voltage sources, the possibility of grounding the electrodes and capacity compensation. Besides these specialized amplifiers, in principle every patch-clamp amplifier can also be used for bilayer experiments. The only restriction sometimes is that patch-clamp amplifiers may be designed for too low an input capacity, due to the small area of patches. As bilayers usually have bigger areas than patches, the range of input capacity of these amplifiers may not be sufficient for the needs of bilayer experiments.

Alternatively, as stated earlier, one can build one's own amplifiers. In Figure 3.21 the principles of a bilayer set-up are given again, focusing on the electronic part. The most simple possible I/V-converter is included in this diagram. It consists of an operational amplifier, a feedback resistor and a power supply. The input resistance of the operational amplifier should be as high as possible ($>10^{12}$ ohm), the feedback resistor should be big enough for the output signal to be in the range of at least 10 mV/pA (10 Gohm) for single-channel experiments (Alvarez *et al.*,

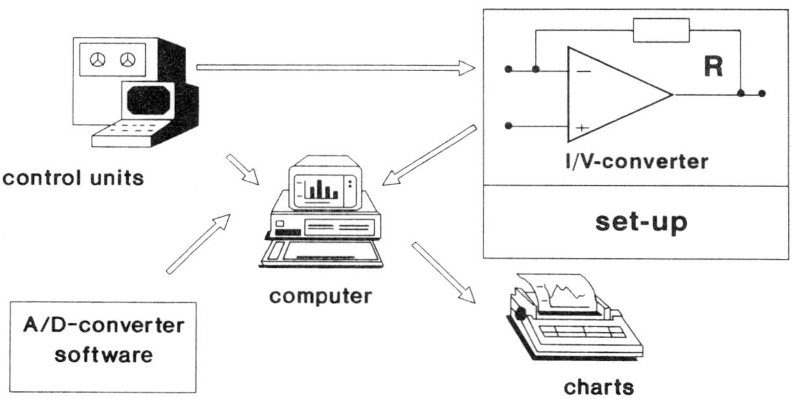


**Figure 3.21** Basic electronics of a bilayer set-up. The main part of the system is a current to voltage (I/V) converter of sufficient sensitivity (input impedance higher than $10^{12}$ ohm). The I/V-converter should be built from a FET-input operational amplifier (high input impedance) and a feedback resistor of about $10^{10}$ ohm.

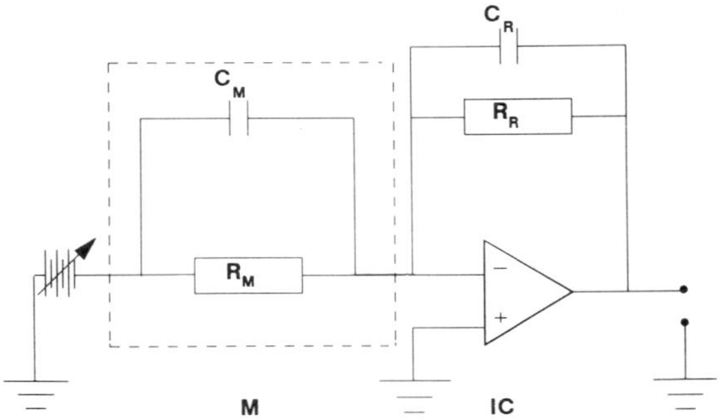


**Figure 3.22** Real I/V-converter to be used in a bilayer set-up, built from a high performance operational amplifier. In addition to the feedback resistor a parallel capacity is drawn; this can not be avoided as any electronic component has a parasitic capacity. Such a capacity will reduce the time resolution of the I/V-converter significantly.

1985), or smaller in case of multi-channel or other high-current experiments. The output signal is calculated according to:

$$V_{out} = I_{in} \times R \qquad [3.1]$$

Sources of operational amplifiers and resistors are given in Appendix 4 (Analog Devices, 1991; Burr Brown, 1991).

The very basic design given in Figure 3.21 is partially misleading. Any resistor and any layout comes together with some capacity, thus a more realistic situation is given in Figure 3.22, here a feedback capacity is included in the layout. This capacity induces a time constant to the amplifier according to:

$$\tau = R \times C \qquad [3.2]$$

Due to the high values which are needed as feedback resistor, ($>10^9$ ohm), and an inevitable capacity in the pico-farad range, this time constant may be in the 100 ms range. Because higher time resolution is required, the feedback capacity must be compensated somehow. A circuit which will do so is shown in Figure 3.23, together with a diagram to demonstrate the frequency response of a compensated and an uncompensated I/V-converter.

In addition to the problem of time resolution, when building an I/V-converter one has to take care with the signal to noise ratio. Time resolution and signal to noise resolution of course are not independent of each other, and as a good basis for calculation one can state that the best possible product of time and current resolution is about $10^{-16}$ A s (improved I/V-converters with a time resolution up to 100 kHz and a good signal to noise ratio are available commercially or can be built according to plans in the literature; see Figure 3.23). With home built amplifiers a time resolution of 10 kHz can be reached without any

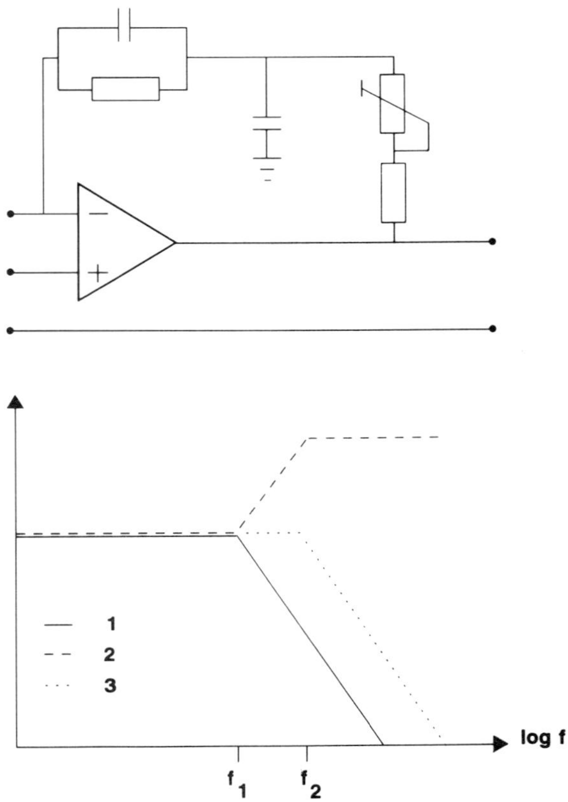


**Figure 3.23** Enhanced I/V-converter design with higher time resolution than that given in Figure 3.22. The capacity of the feedback circuit is compensated by a proper circuit layout. A simplified diagram of the frequency reponse of the compensated and the non-compensated I/V-converters is added below the circuit diagram. The operational amplifier used, should be a FET-input type (field effect transistor input stage) of high input resistance and low bias current (for example the Burr-Brown OPA-128).

problem *or* a current resolution of about 0.1 pA.

### 3.3.2 Command-voltage supply

A command voltage must be supplied to a bilayer, in order to measure current across it. This can be done either by the built-in voltage source of commercially available amplifiers or by a simple device as shown in Figure 3.24 (upper). This device is made from two batteries, two resistors and a potentiometer only. In the lower part of Figure 3.24 a layout which is slightly more sophisticated is given. An operational amplifier is used with different supplies for command voltage and adjustment; it includes an input for a function generator to add external signals to the membrane potential. Additionally, a source to compensate electrode potentials is included. This type of command voltage source will suffice for most experiments.

### 3.3.3 Capacity measuring devices

One basic important electrical property of a bilayer is its capacity. It can be measured in a very simple way, as shown in Figure 3.25. A voltage pulse is applied to the bilayer via a known series resistor. The time-course of the resulting current is measured. From the decay time of this signal the capacity can directly be calculated from equation 3.2. A better and continous way to record the membrane capacity is shown in Figure 3.26. To the bilayer a triangular-wave is applied. According to the current across a capacity as response to a time dependent voltage:

$$I = C \times dV/dt \qquad [3.3]$$

the capacity can be calculated from the amplitude of the resulting square wave. It is best to calibrate the system with capacitors of known value.

In an even more sophisticated set-up a lock-in amplifier can be used to measure the bilayer capacity continously. Such devices have been

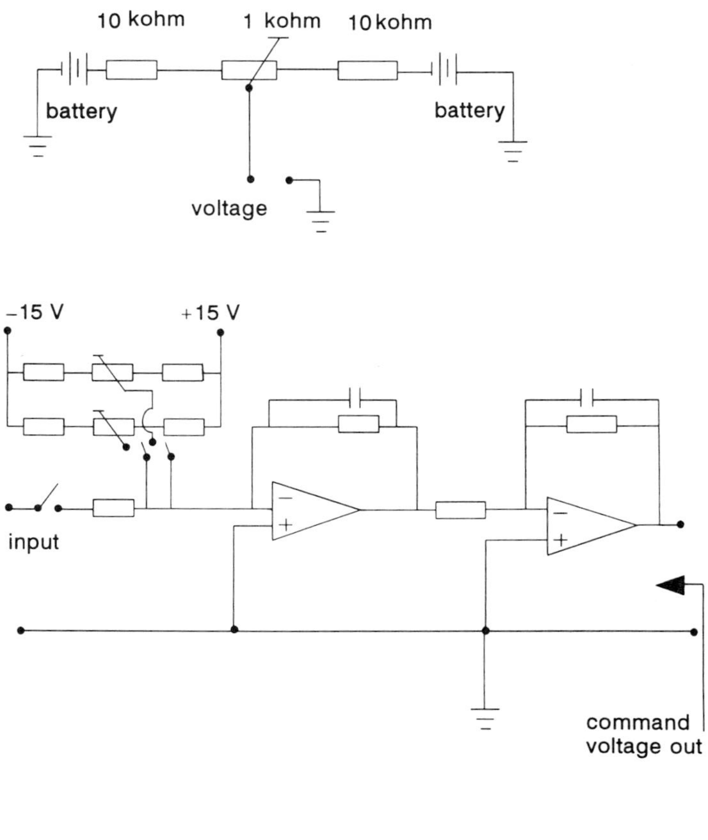


**Figure 3.24** Upper: the design of a really simple command voltage supply is shown. Lower: a more sophisticated design, which will fit most of the needs of bilayer experiments. The potential range usually required in bilayer experiments ranges from some +500 mV to some −500 mV, the resistors in the circuit must be set accordingly.

used in patch-clamp experiments for high time and capacity resolution (Neher and Marty, 1982).

Instead of using a high-price lock-in amplifier, in a simplified set-up a sine-wave can be added to the bilayer and the amplitude of the resulting cos-wave (current) can be taken as a measure of the bilayer capacity. When the frequency of the sine-wave used in such devices is very high a low-pass filter with a steep cut-off at a frequency about two times lower than the command signal frequency can be used and capacity and DC current can be measured independently. In Figure 3.27 a diagram of such a system is given.

A variety of other devices which can be used to measure bilayer capacity have been described (Lindau, 1991; Toyama *et al.*, 1991; Kalinowski and Figaszewki, 1992).

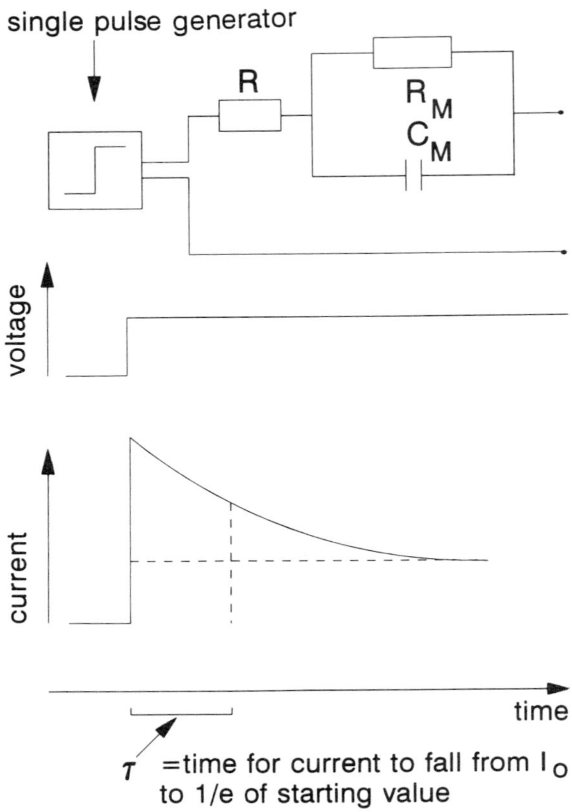


**Figure 3.25** Measurement of bilayer capacity by use of a single pulse supplied over a high series resistor to the membrane. The time constant ($\tau$) for the transient signal to fall from $I_o$ to 1/e from any starting point can be used to calculate the bilayer capacity (Equation 3.2).

### 3.3.4 Recording devices

Usually electrical signals from bilayer experiments have to be recorded and stored for subsequent analysis. In most bilayer laboratories today signals are displayed on an oscilloscope during the experiment. Parallel to this, slow signals (up to 100 Hz) may be recorded directly on high-speed chart recorders. It is better, however, to use a tape recorder for mass storage. Two types of such recorders are available:

- using FM-modulation to record the signal from DC up to some 10 kHz
- using PCM-modulation (same frequency response).

The advantage of FM tape recorders is that the signal can be replayed at slower speed to record it directly on a strip-chart recorder. The signal to noise ratio of these FM recorders, however, is low (about 48 dB). PCM recorders have a much better signal to noise ratio (>80 dB) but usually can only be used at one speed. Today some PCM machines are available with direct access to computers. From these machines the data can be directly read by a computer for further data evaluation.

## 3.4 DATA ACQUISITION AND COMPUTER SUPPORT

As computers are present in nearly every laboratory today, no-one would try to evaluate bilayer data by hand but instead would use a computer. Especially with single-channel data this is of evident advantage.

To evaluate bilayer data by a computer an interface for data aquisition must be available as well as appropriate software. Data can be recorded by the computer directly during the

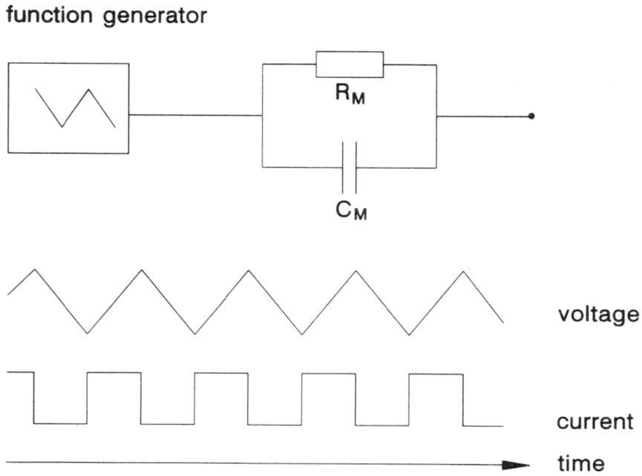


**Figure 3.26** Continuous recording of bilayer capacity by use of a triangular wave applied to the bilayer. The amplitude of the resulting square wave (in the current recording) is a measure of the bilayer capacity ($I = C^{*}\mathrm{d}U/\mathrm{d}t$). It is best to calibrate the system before use with capacitors of known value.

experiment or later from a tape recorder when data have been recorded on such a device during the experiment. In both cases an analog to digital converter (A/D-converter) is needed to digitize the data. After that, appropriate software can be used for further data evaluation. Such software (limited to single-channel current recordings) is described in Chapter 7 in detail. Of course all other parameters measured in a bilayer experiment can be recorded on a computer too (pH, temperature, potentials). There are no real problems with the A/D-converter as the technical requirements are not very high. Sampling rates of a few tens of kHz (up to 100 kHz) should be possible at a minimum of 12-bit amplitude resolution. Two channels enabling at least parallel recording of potential and current are adequate. The sampling rate adjusted by the software in computer-controlled measurements

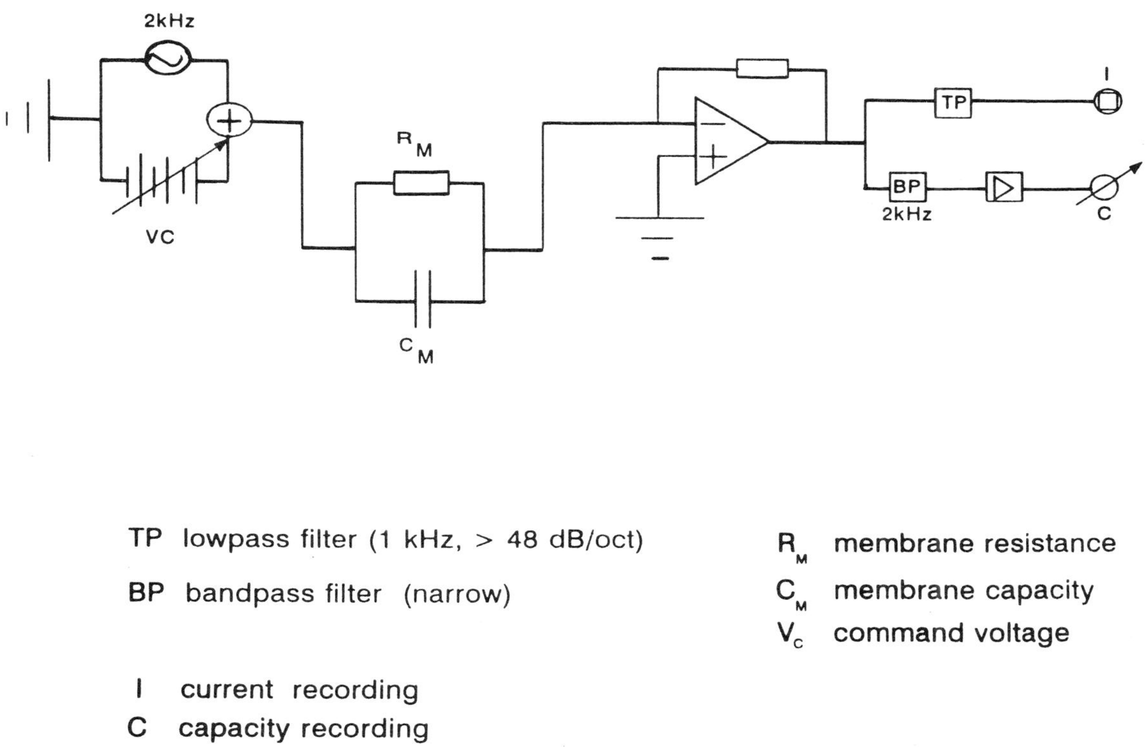


**Figure 3.27** System for continuous recording of bilayer capacity during measurement of ion channel current fluctuations. A sine wave of sufficiently high frequency (2 kHz) is applied to the bilayer. The ion channel current fluctuations are measured in the frequency band of 0 Hz (DC) up to 1 kHz using a low pass filter with a sharp cut-off at 1 kHz (Bessel type, at least 48 dB per octave). The cosine wave resulting from the sine wave command signal is measured in parallel, using a band-pass filter at 2 kHz (or lock in amplifier tuned properly).

should be at least twice (better if four times) the highest frequency one wants to resolve in the data (digitizing theorem). A/D-converters with the analogue circuit placed outside the computer itself have the advantage of being less sensitive to electrical distortions from the power supply of the computer (and the digital circuit) and are therefore recommended.

## 3.5 ADDITIONAL EQUIPMENT

In specialized bilayer experiments other than those described above, other technical components may be necessary. We will give a list here of some pieces of equipment which might be practical but in simpler applications are usually not required:

1. D/A-converter for computer-controlled voltage protocol
2. sophisticated function generators
3. additional filters
4. electrodes other than Ag/Ag-Cl electrodes (see also section 3.1.3).

Of course we have not listed all general items in the preceding chapters, such as oscilloscopes, pen recorders or simple function generators. The above list can not be complete in any case. After having built a first basic set-up, each experimenter will find his own needs with time. In Chapter 8 we will describe some special applications together with more sophisticated equipment for those with some experience in the field.

## REFERENCES


Analog Devices (1991) *Integrated circuits data book*. Analog Devices, Norwood, Massachusetts.

Alvarez, O., Benos, D. and Latorre, R. (1985) The study of ionic channels in planar lipid bilayer membranes. *J. Electrophysiol. Tech.* **12**, 159–178.

Burr-Brown (1991) *Integrated circuits data book*. Burr-Brown, Tucson, Arizona.

Coronado, R. (1986) Recent advances in planar phospholipid bilayer techniques for monitoring ion channels. *Ann. Rev. Biophys. Chem.* **15**, 259–277.

Corvington, A.K. (1979) *Ion-selective electrode methology*, Vol. I. CRC Press, Boca Raton, Florida.

DuBovy, J. (1978) *Introduction to biomedical electronics*. McGraw-Hill, New York.

Fishman, H.M. (1982) Current and voltage-clamp techniques. In *Techniques in cellular physiology*. Elsevier/North Holland, Country Clare, Ireland.

Hamill, O.P., Marty, A., Neher, E., Sakmann, B. and Sigworth, F.J. (1981) Improved patch-clamp techniques for high-resolution current recording from cells and cell-free membrane patches. *Pflügers Arch.* **391**, 85–100.

Haughland, R.P. (1992) *Molecular probes: Handbook of fluorescent probes and research chemicals*. Molecular Probes Inc., Eugene.

Inoue, S. (1986) *Video microscopy*. Plenum Press, New York.

Kalinowski, S. and Figazewski, Z. (1992) A new system for bilayer lipid membrane capacitance measurements: method, apparatus and application. *Biochim. Biophys. Acta* **1112**, 57–66.

Kirino, Y., Hirashima, N. and Saito, Y. (1991) Patch-clamp measurements of ion channels in biomembrane vesicles reconstituted into giant proteoliposomes by freeze-thawing without use of detergence. In *Electrophysiology and microinjection*. P.M. Conn (ed.) Academic Press, San Diego.

Latorre, R. (1986) *Ionic channels in cells and model systems*. Plenum Press, New York.

Lindau, L. (1991) Time-resolved capacity measurement: Monitoring exocytosis in single cells. *Quart. Rev. Biophys.* **24**, 75–101.

Miller, C. (1986) *Ion channel reconstitution*. Plenum Press, New York.

Neher, E. and Marty, A. (1982) Discrete changes of cell membrane capacitance observed under conditiones of enhanced secretion in bovine adrenal chromaffin cells. *Proc. Natl. Acad. Sci. USA* **79**, 6712–6716.

Niles, W.D. and Cohen, F.S. (1991) Fusion of influenza virions with a planar lipid membrane detected by video fluorescence microscopy. *J. Gen. Physiol.* **97**, 1101–1119.

Russ, J.C. (1990) *Computer-assisted microscopy. The measurement and analysis of images*. Plenum Press, New York.

Sakmann, B. and Neher, E. (1983) *Single-channel recording*. Plenum Press, New York.

Slayter, E.M. (1970) *Optical methods in biology*. Wiley-Interscience, New York.

Thomas, R.C. (1978) *Biological Techniques*. Ion-sensitive electrodes. Academic Press, London.

Ti Tien, H. (1974) *Bilayer lipid membranes: Theory and practice*. Marcel Dekker, New York.

Toyama, S., Nakamura, A. and Toda, F. (1991) Measurement of voltage dependence of capacitance of planar lipid bilayer membrane with a patch clamp amplifier. *Biophys. J.* **59**, 939–944.

Wilson, L. (series ed.) (1989) *Methods in cell biology*, Vol. 30. *Fluorescence microscopy of living cells in culture*. Part A: Quantitative fluorescence microscopy – imaging and spectroscopy. Academic Press, London.

Zimmerberg, J., Cohen, F.S. and Finkelstein, A. (1980) Fusion of phospholipid vesicles with planar lipid bilayer membranes: I. Discharge of vesicular content across the planar membrane. *J. Gen. Physiol.* **75**, 241–250.

CHAPTER FOUR

# Biochemical preparations for planar lipid bilayer experiments

As stated in the previous chapters, biological membranes are highly complicated systems composed of a variety of biochemical components (Jain and Wagner, 1980). Some data summarizing the composition of typical biological membranes are given in table 4.1. The main components of these membranes are lipids and proteins: the detailed composition of biological membranes, however, is very often not known or may even vary from cell to cell within identical material, a fact making *in vivo* experiment difficult as far as questions about the influence of membrane composition on transport systems are concerned.

**Table 4.1** Gross chemical composition (% by weight) of some biological membranes (data from the literature vary from author to author).

| *Membrane type* | *Protein* | *Lipid* | *Carbohydrate* |
|---|---|---|---|
| Myelin | 18 | 78 | 4 |
| Plasma membrane | | | |
| Human erythrocyte | 49 | 43 | 8 |
| Mouse liver | 45 | 52 | 3 |
| Amoeba | 54 | 42 | 4 |
| Mitochondria | | | |
| Inner membrane | 76 | 24 | 0 |
| Chloroplast | 70 | 30 | 0 |
| Rod outer segments | 60 | 40 | 0 |

One of the advantages of bilayer experiments is the possibility of creating membrane systems with well-defined and known biochemical composition. This makes it necessary, however, to have adequate preparations of lipids and proteins at hand. In this chapter we offer some general remarks on biochemical and molecular membrane structure of membrane compounds and introduce techniques for preparing lipids and proteins for use in bilayer experiments.

Of course lipid mixtures purified from natural membranes as well as highly purified or synthetic lipids are commercially available (a list of some selected sources is given in Appendix 4). Nevertheless, it is useful to be able to prepare or investigate lipids in the laboratory by oneself. Some tips for this are given in this chapter.

In the case of proteins and polypeptides, preparations are often not available commercially.

Proteins, usually mediating membrane transport, have to be prepared according to the needs of the particular experiment; a few examples are given in this chapter. As usual, some exceptions do exist; simpler pore-forming polypeptides and carriers are commercially available, examples are:

1. gramicidin
2. alamethicin

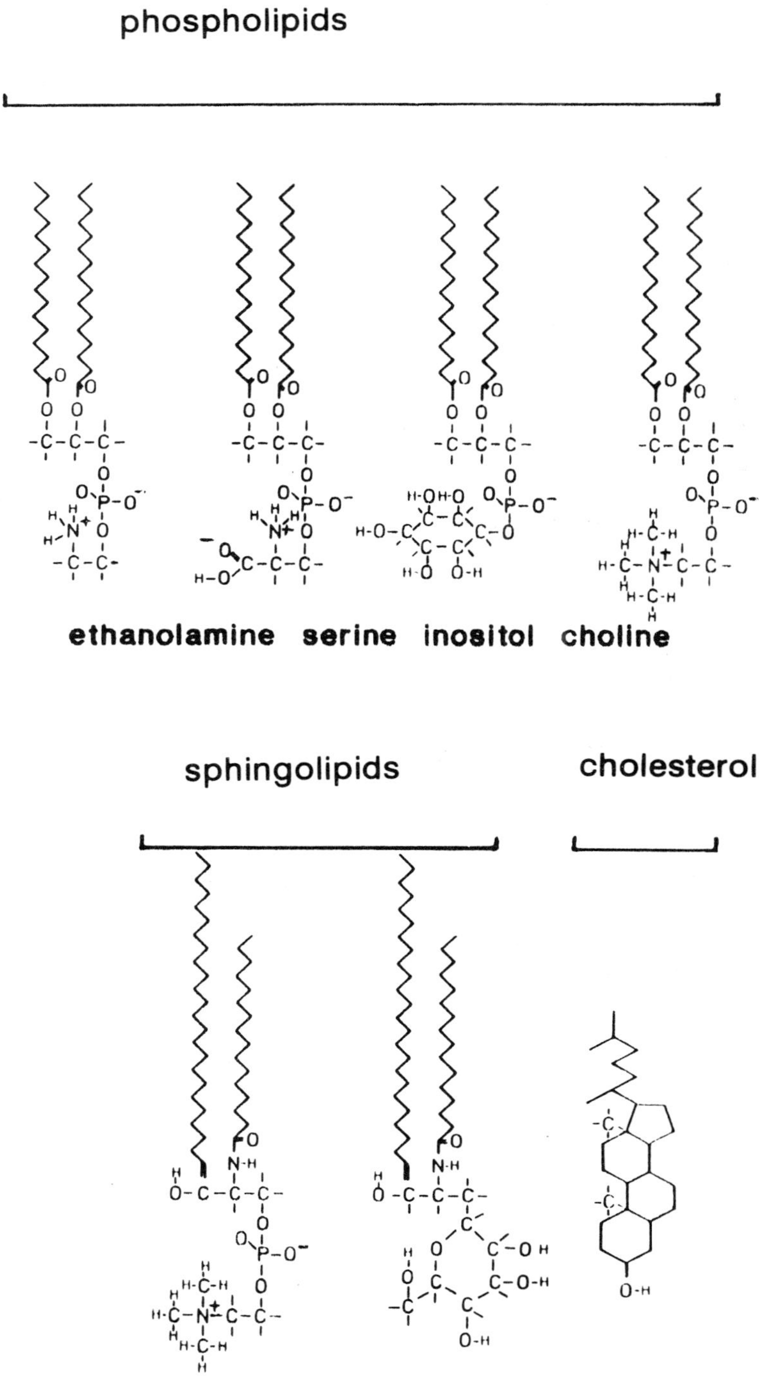


**Figure 4.1** Chemical composition, structure and classification of some important lipids found in natural membranes. The structure of the hydrocarbon chains is not given in detail; see Table 4.2 for additional information.

**Table 4.2** Gross lipid composition (% by weight) of some selected membranes (data from Jain and Wagner, 1980).

| *Membrane* | *Cholesterol* | *PC* | *SM* | *PE* | *PI* | *PS* | *PG* | *DPG* | *PA* | *Glycolipids* |
|---|---|---|---|---|---|---|---|---|---|---|
| Rat liver | | | | | | | | | | |
| Cytoplasmic membrane | 30 | 18 | 14 | 11 | 4 | 9 | | | 1 | |
| Endoplasmic reticulum | | | | | | | | | | |
| Rough | 6 | 55 | 3 | 16 | 8 | 3 | | | | |
| Smooth | 10 | 55 | 12 | 21 | 7 | | | 2 | | |
| Myelin | 22 | 11 | 6 | 14 | | 7 | | | | 12 |
| *Escherichia coli* | 0 | 0 | | 80 | | | 15 | 5 | | |
| Rat erythrocyte | 24 | 31 | 9 | 15 | 2 | 7 | | | | 3 |

Abbreviations: PC, phosphatidylcholine; SM, sphingomyelin; PE, phosphatidylethanolamine; PI, phosphatidylinositol; PS, phosphatidylserine; PG, phosphatidylglycerol; DPG, diphosphatidylglycerol; PA, phosphatidic acid.

3. valinomycin
4. nigericin
5. different types of ionophores (especially for calcium).

Again the sources are given in Appendix 4. Therefore we will not discuss the preparation of such material but will focus on the preparation of transport proteins from biological membranes to various degrees of purity (Section 4.2).

## 4.1 LIPIDS

### 4.1.1 Characterization of lipids

Lipids make up the core of any biological membrane in the form of a bilayer; as seen from Table 4.1 their percentage may range from 10–80% of total membrane material. Because of the variety of biological tasks that membranes have to fulfil, an enormous number of different lipids is found in nature. Usually these lipids are classified according to their hydrophilic head-groups as in Table 4.2. A textbook classification of the more important classes of lipids is given in Figure 4.1 together with drawings of their principal structures. As can be seen, the head-groups may differ in their size as well as in the electrical charge they carry. In addition to this classification, lipids may be distinguished according to the nature of their hydrocarbon chains. Some criteria which may help in a detailed classification of lipids by the structure of their hydrophobic chains are:

1. alcohol, fatty acid or other compound
2. length of hydrocarbon chain
3. number and position of double bonds.
4. existence of triple bonds.

In tables 4.2 and 4.3 a summary is given of the lipid composition of some typical biological membranes and some more detailed information about these lipids (for a classification of lipids see for example Kates, 1986).

**Table 4.3** Fatty acid composition (% of membrane content) of phospholipids in human erythrocytes.

| *Fatty acid* | *Length* | *Double bonds* | *Content* |
|---|---|---|---|
| Palmitic | 16 | 0 | 35 |
| Stearic | 18 | 0 | 13 |
| Oleic | 18 | 1 | 19 |
| Linoleic | 18 | 2 | 21 |
| Arachidonic | 20 | 4 | 5 |
| Others | | | 7 |

### 4.1.1 Purification of lipids

For standard bilayer experiments, natural lipids are often prepared from biological membranes in different degrees of purity. A typical lipid of this group which is widely used in bilayer reconstitution experiments is Asolectin, a lipid extract from soya bean membranes. It is commercially available at very low price and contains some 10% of lecithin. Its detailed composition is not standardized and therefore a rough characterization of each batch is useful. Additionally it may contain up to some 5% proteolipids and other contaminants. Due to the low price of the material it is not worth preparing it in the laboratory. As stated, however, a test of the material is recommended (for example by thin-layer chromatography) and a partial purification (to get rid of the proteolipids) of the Asolectin is often done before use.

The easiest test of the quality and composition of a lipid is to do a thin-layer chromatography (TLC) on the material. Details of TLC techniques are given in text books about lipid technology and TLC itself (Stahl, 1969; Bergelson, 1980; Kates, 1986); here we will describe a fast screening and test procedure using Asolectin as an example (using commercially available equipment).

1. Dissolve Asolectin in chloroform:methanol, 2:1 (v/v) 4–5% solution.
2. Apply 0.5–2.0 µl of solution to a standard analytical TLC plate (silica gel) which has been heated before use. Use a syringe, for example, a Hamilton syringe.
3. Let plate dry.
4. Put plate into glass tank of appropriate size (10 × 30 × 20 cm, $d \times w \times h$) , filled with solvent (60 ml) (chloroform: methanol: acetic acid: water, 65: 43: 1: 3). The tank should be lined with filter-paper to aid in saturating the tank with vapour.
5. Close tank with glass plate and wait until solvent front is close to the upper border of the TLC plate.
6. Take out TLC plate and dry under nitrogen.
7. Spray plate with rhodamine 6G (commercially available).
8. View plate wet under UV-light (366 nm) to see yellow or blue spots.
9. Mark spots with pencil on the plate.

Instead of points 7 and 8 the plate can be placed in a container with iodine vapour; yellow and brown spots will result.

The TLC plates are best used with two lines of the lipid (outer lines) to be investigated and additional lines of lipid standards (lecithin, kephalin, phosphatidylserine, phosphatidylinositol and others) for comparison. Numerous different procedures for TLC techniques have been described, and it may be necessary to modify the procedure described above in certain cases (see the literature about lipid technology: Bergelson, 1980; Kates, 1986).

After having analysed the composition of a lipid mixture, here Asolectin, a partial purification may be required as stated already. A simple procedure to remove proteolipids from a lipid mixture and to concentrate its phospholipid content is acetone precipitation. This procedure makes use of the fact that phospholipids are not soluble in cold acetone. The procedure is as follows:

1. Dissolve 10 g of lipid in 100 ml chloroform: methanol, 1:2, (v/v).
2. Add 20 ml of 3 M NaCl and homogenize.
3. Let phases separate in a separation funnel (about 1 h).
4. Take chloroform phase and dry it in a rotovac.
5. Dissolve material in chloroform (50 ml).
6. Add 250 ml of acetone (ice cold) and stir for about 1 h in a cold room.
7. Save precipitate (spin down at 2500 rpm for a few minutes in a centrifuge, using glass tubes if possible).

The complete procedure must be carried out in a cold room (4°C). The pellet will be concentrated mainly in neutral phospholipids. For a very fast even more simplified procedure, points 1 to 4 may be omitted. Additionally, in point 2, to the aqueous solution 10 mM ethylene glycol tetra-acetic acid can be added to remove all calcium from the lipid, a procedure sometimes required when measurements in calcium-free solutions must be done. It may be useful to check the material again by TLC for comparison with the starting material after the purification procedure. The material obtained by this technique from crude Asolectin (or partially purified Asolectin) can be used in bilayer experiments for reconstitution of proteins with good success.

Sometimes pure lecithin or kephalin (or another lipid) is required for certain experiments. Some of these pure lipids can be prepared from partially pre-purified Asolectin (see above) or, for example, from egg yolk, which is a good starting material to purify lecithin and kephalin. We will give here a procedure for the purification of lecithin and kephalin from egg yolk, as these lipids are widely used in bilayer reconstitution experiments:

1. Separate egg yolk from 10 eggs and determine weight.
2. Homogenize material in 0.3 vol of 2 M NaCl and 2 vol of methanol.
3. Add 1 vol of chloroform and homogenize again.
4. Spin down in glass bottles for 20 min at 4000 rpm (about 2000 *g*).
5. Save supernatant.
6. Add 1 vol. of water and 1 vol. of chloroform and homogenize.
7. Let phases separate in funnel and take the lower phase.
8. Dry extract to about 100 ml in a rotovac.
9. Add 500 ml of cold acetone, stir for 1 h.
10. Save precipitate in 20–40 ml of chloroform.

At the end of this part of the procedure you will have a crude preparation of mainly phospholipids and some other lipids from egg yolk, this lipid mixture can be used directly for bilayer experiments if high purity of the lipids is not required.

The material can now be separated to obtain pure phospholipids by column chromatography on a silica gel. Of course, one can prepare the columns in the laboratory, but pre-packed ready-to-use silica-gel columns for low pressure liquid chromatography are available at moderate prices (for example, MERCK low-bar columns). Prepare such a column as described in manuals or in the literature.

1. Equilibrate column in chloroform until it is clear.
2. Add lipid in chloroform (use a column size according to the amount of material which must be purified). As starting material, the pre-purified Asolectin preparation can be used.
3. Run in chloroform until first yellow band comes out.
4. Start collecting fractions of about 150 ml and run chloroform:methanol (C:M, 4:1) until the second yellow band comes out.
5. Start collecting fractions of about 10 ml and switch to C:M, 3:2.
6. Kephalin (PE) will appear at around 250 ml, depending of course on column size and amount of starting material (test every fraction on TLC).
7. Pool PE fractions.
8. Switch to methanol and wait for lecithin (PC); collect 50 ml fractions and screen with TLC.
9. Pool PC fractions.

The pooled PC and PE fraction can be dried in a rotovac and stored under liquid nitrogen until use. It is recommended that the final material is tested on TLC for purity. Sometimes impurities are found, which can be removed in a second chromatographic step on a diethylaminoethanol (DEAE) column (Kates, 1986).

The lipid preparations described here serve as examples of what can be done in a moderately equipped laboratory. As stated already, all types of lipids are available commercially. When it is required that lipids are not only pure in head-group but also in hydrocarbon-chain composition, more sophisticated biochemical methods are necessary to prepare them. Usually this can not be done in a bilayer laboratory, thus these lipids should be bought.

Instead of using low pressure chromatography, of course, lipids can be purified by high performance liquid chromatography (HPLC) if the necessary equipment is available. As this equipment is very expensive and usually not present in bilayer laboratories, we will not focus on the HPLC techniques here.

Biochemical procedures have been developed to change the head-group of lipids or to modify the hydrocarbon chain composition. All these techniques require deeper knowledge of biochemistry and we recommend the interested reader to the specialized literature.

### 4.1.3 Synthetic lipids

An alternative to the use of purified lipids in bilayer experiments is to use synthetic lipids which are also available commercially. These lipids are pure in their head-group as well as in their hydrocarbon chain composition. It is additionally possible to synthesize lipids which are not present in biological membranes. The use of such lipids may sometimes be of interest in special biophysical experiments. Some of the artificial lipids used in bilayer experiments are:

1. positively charged lipids
2. lipids which can be polymerized with UV-light

(due to the existence of triple bonds in the hydrocarbon chains)

3. phospholipids with the hydrocarbon chains in the 1 and the 3 position of the glycerol

Of course a lot of other structures can be created. All these lipids must usually be bought as synthesis in an electrophysiological laboratory seems to be quite difficult. However, some of them are not even commercially available and can only obtained from laboratories specialized in lipid biochemistry.

## 4.2 PROTEINS

Biological membranes contain significant portions of proteins (see Table 4.1) which are thought to be responsible (besides other functions) for the transport properties of a membrane. The reconstitution of such transport proteins into artificial membranes (planar lipid bilayers) is one of the major groups of experiments in bilayer laboratories. To make proteins available for integration into artificial membranes, appropriate preparations of the proteins must be made. In this chapter some procedures are outlined to prepare proteins from biological membranes for use in bilayer experiments. Some principles of such preparations have already been given in Chapter 1 (Figure 1.2). As stated there, proteins can be prepared in different degrees of purity, starting from crude membrane preparations and proceeding to highly purified proteins or even single polypeptides or subunits of proteins.

### 4.2.1 Membrane fragments

The most simple method to incorporate a protein from a biological membrane into a planar lipid bilayer is to use fragments of the membrane from which the protein originates. In this case, only the membrane fragments have to be prepared and no further purification is required. In fact most proteins reconstituted to date in bilayers have been first reconstituted from such membrane fragments. In the following we will give a procedure for making membrane fragments from *Torpedo* electroplax, a tissue source material used for reconstitution of acetylcholine receptors and chloride channels (Sobel *et al.*, 1977; Hanke and Miller, 1983). The major task of such a procedure is to obtain a preparation which only contains that type of membrane the protein has been shown to be a constituent of:

1. Take electric organ of *Torpedo californica*.
2. Cut off connective tissue.
3. Cut organ into small pieces.
4. Add 1 vol. of 350 mM NaCl, 50 mM KCl, 10 mM $NaHPO_4$, 10 mM EGTA, 5 mM phenylmethylsulfonyl fluoride (PMSF), pH = 7.0, 5 mM indole-3-acetic acid (IAA).
5. Homogenize in Waring blender (3 min).
6. Homogenize in Vortex blender (30 sec).
7. Spin down 10 min at 5000 rpm (about 3000 *g*).
8. Take supernatant, filter.
9. Spin down 60 min at 25 000 rpm (about 60 000 *g*).
10. Collect pellets in solution containing 10 mM $NaHPO_4$, 5 mM EGTA, pH = 7.4; try to avoid mitochondria which collect as a dark spot below the bulk of the pellet.
11. Homogenize in glass-homogenizer, measure protein content, spin down at 5000 rpm (about 3000 *g*) for 10 min and throw away pellet.
12. Make sucrose gradient solutions for centrifugation 55 % (w/w), 39 % (w/w), 35 % (w/w).
13. Add sucrose to supernatant from point 11, up to 30 % (w/w) with 350 mM NaCl, 50 mm KCl.
14. Make sucrose gradient in appropriate tubes.
15. Use vertical rotor, 45 000 rpm (about 150 000–200 000 *g*) for 90 min, for reorienting gradient, or spin overnight in swing out rotor.
16. Collect protein bands as visualized by scattered light and add 1 vol distilled water.
17. Spin down for 60 min at 25 000 rpm (about 60 000 *g*) (or for 30 min at 35 000 rpm).
18. Collect pellets in 200 mM sucrose, 10 mM $NaHPO_4$, pH = 7.
19. Store at −70°C.

The procedure outlined here does not give all details of the preparation, but it should be sufficient to demonstrate the principles of a membrane fragment preparation. When making preparations it is recommended that the original literature on such procedures is referred to.

The example of preparing membrane fragments from *Torpedo* electrocytes is just one from numerous examples described in the literature to make membrane fragment preparations for bilayer reconstitution experiments (it was used by Miller and coworkers – (Tank *et al.*, 1982; Hanke and Miller, 1983) – especially to obtain membranes which have high levels of chloride channels). Some other typical membrane fragment preparations are:

1. Sarcoplasmic reticulum (SR)-preparation to reconstitute $K^+$-channels (Labarca *et al.*, 1980)
2. rod outer segment plasma membrane preparation to reconstitute cGMP-activated channels (Cook *et al.*, 1986, 1987)
3. synaptosomes to reconstitute $Ca^{2+}$ and other channels (Tareilus *et al.*, 1990)
4. bacterial membranes to reconstitute porins (Benz, 1985).

Each of the preparations then can be used together with different bilayer techniques and reconstitution techniques as described in the following chapters (for further information see also: Sakmann and Neher, 1983; Latorre, 1986; Miller, 1986; Findlay and Evans, 1987).

### 4.2.2 Protein purification

Even more tempting than using membrane fragment preparations for reconstitution is the use of 'pure' proteins in lipid bilayers, as discussed in more detail in the introduction. We want to limit ourselves in this description to techniques of purifying integral membrane proteins. A major problem of protein purification is that the proteins must be removed from their natural membrane environment during purification. After doing that, one type of protein, which by some means has been identified before in its natural membrane, must be separated from all the other proteins from the membrane. Finally, the protein purification procedure should leave the functional properties of the proteins unchanged; usually this must be verified by additional experiments. Different techniques have been developed to fulfil all the requirements stated here.

Before going to a more detailed description of specialized techniques, some general remarks on protein purification procedures are made. These procedures can be separated into two main classes:

1. a functional purification, making use of a known function of the protein to identify and purify it (an example of this technique will be given later)
2. a purification according to the specific binding of a drug to the protein (affinity chromatography); again, an example will be given later.

In Figure 4.2 a general scheme is given of a functional purification procedure. One very well-known example of such a preparation is the cGMP-activated channel from bovine rod outer segment membranes as developed by Kaupp and collegues (Cook *et al.*, 1986, 1987). This channel was found to be permeable to cations, including $Ca^{2+}$. The $Ca^{2+}$ permeability of the intact channel, together with its activation by cGMP, was used as an assay for channel purification. In Figure 4.3 a description of this special functional assay is given. The cGMP-triggered $Ca^{2+}$ flow out of $Ca^{2+}$-loaded vesicles is measured using vesicles that contain the purified protein in their membranes. We will not give a detailed description of the purification procedure here, as it requires deeper biochemical knowledge, but instead we show in Figure 4.4 a flow chart of the procedure. This scheme includes the following steps:

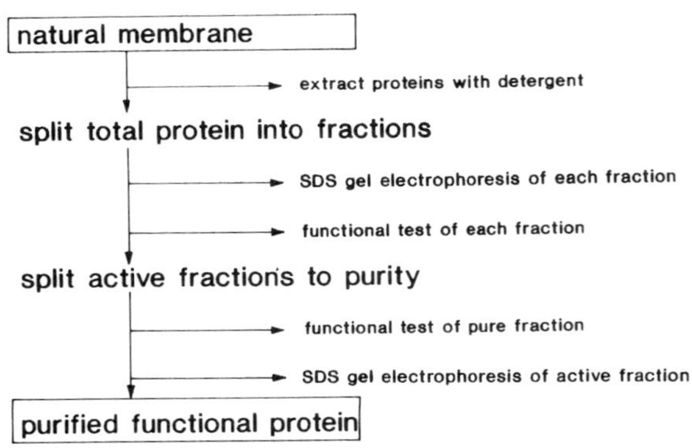


**Figure 4.2** Scheme of a purification procedure to obtain a functional protein. Any protein function that can be tested by an assay can be used.

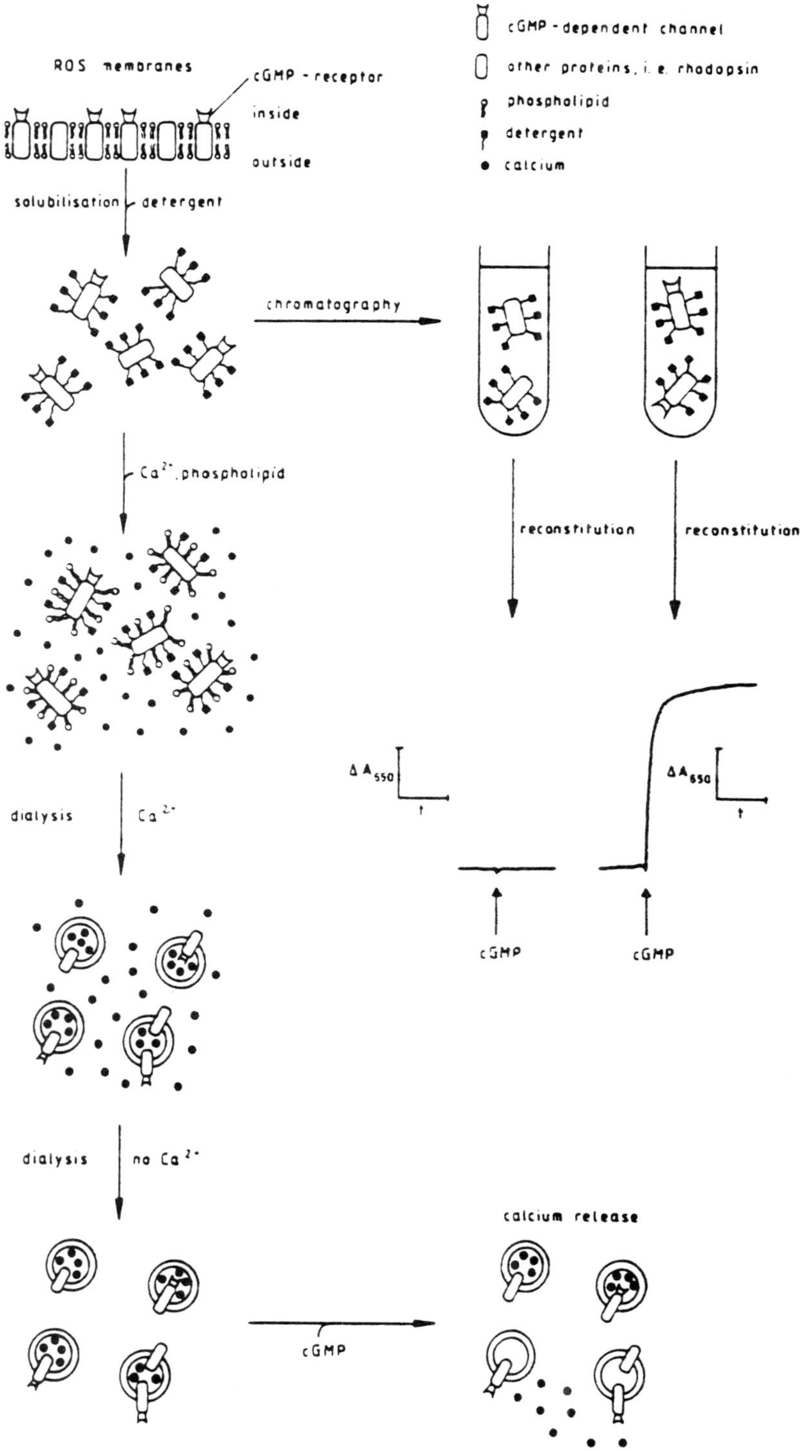


**Figure 4.3** $Ca^{2+}$-efflux assay as used for measuring the functional integrity of cGMP-activated channel from rod outer segment (ROS) membranes. Arsenazo III is used as dye to trace the efflux of $Ca^{2+}$ out of $Ca^{2+}$-preloaded liposomes, which have the protein in their membranes. This technique makes use of the small but significant $Ca^{2+}$-permeability of the cGMP-activated channel and of the fact that the channel is closed when no cGMP is present. $\Delta A_{650}$, light absorption measured in a spectrophotometer at 650 nm; t, time. (Reproduced from Kaupp *et al.*, 1988 with permission.)

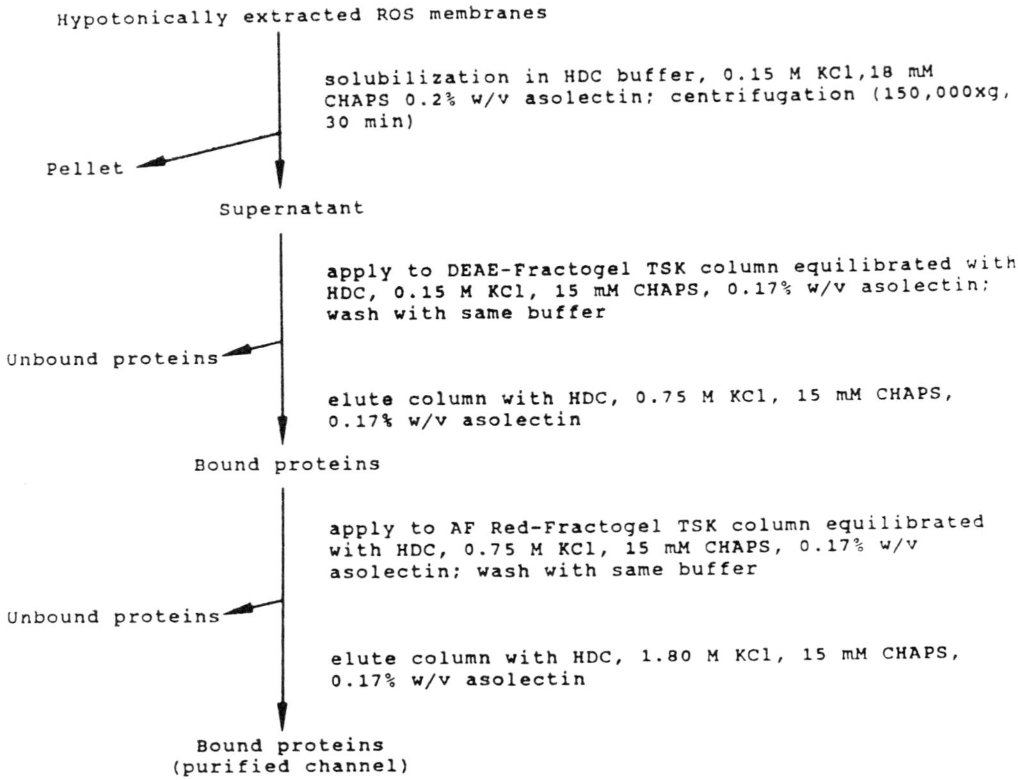


**Figure 4.4** Details of the procedure to purify the cGMP-activated channel from bovine rod outer segments using the technique shown in Figure 4.3 as functional assay. Abbreviations used are: CHAPS, (3–[(3–Cholamidopropyl)-dimethylammonio]-propane-sulfonate); HDC, (HEPES-DTT-$Ca^{2+}$). (Reproduced from Kaupp *et al.*, 1988 with permission).

1. preparation of rods from bovine eyes
2. preparation of the plasma membrane from the rods
3. solubilization of the total protein from the membrane by detergent
4. separation of proteins by a first column chromatography step
5. test of each fraction by a functional assay as described in Figure 4.3
6. second column chromatography step of pooled active fractions from step 4.
7. repeat functional test of each fraction
8. SDS-gel electrophoresis of pooled active fractions to test the purity of the preparation
9. removal of detergents and incorporation of protein into liposome membranes.

In fact, it has been shown that this procedure results in a preparation that only contains one polypeptide band on the SDS-gel, verifying that a pure (most probably homo oligomeric) protein is the result. In Figure 4.5 a SDS-gel of the starting material and the final material are shown.

Other proteins have been also purified using functional assays. Each of the procedures described in the original papers is complex because many fractions obtained with biochemical techniques must be tested for a known function of the protein. The functional test of the fractions can be different according to what is known about the protein to be purified. Within the context of this book it is obvious, furthermore, that a bilayer reconstitution experiment may itself serve as functional test and therefore we will describe such an experiment in more detail in Section 8.6.

The second possibility for purifying a protein is to make use of the specific binding of a known

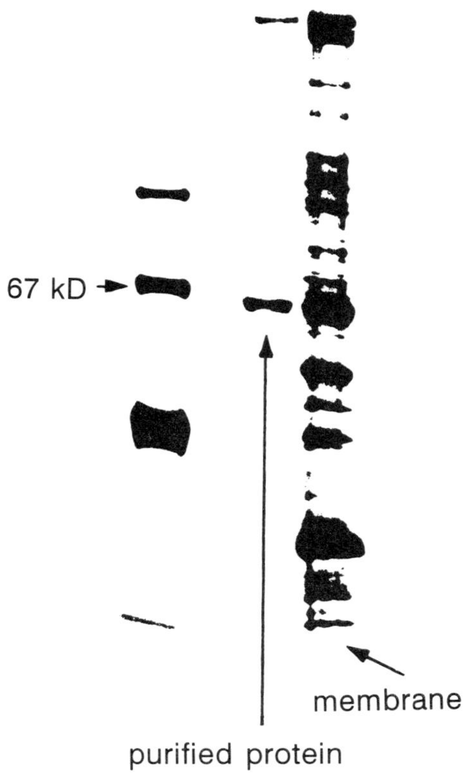


**Figure 4.5** SDS gel of total protein of the rod outer segment membrane (starting material of the purification procedure from Figure 4.4) and of the purified cGMP-activated channel obtained following the procedure given in Figure 4.4. The channel is obviously a homo oligomeric protein (most probably a pentamer) assembled from a 65 kD polypeptide.

**Table 4.4** Ion channels that have been purified by affinity chromatography together with the drugs used.

| *Ion channel* | *Drug* |
|---|---|
| Nicotinic acetylcholine receptor | α-Bungarotoxin |
| $Na^+$ channel | Tetrodotoxin |
| | Saxitoxin |
| $Ca^{2+}$ channel | Dihydropyridines |

drug to the protein. It is known, for example, that specific drugs bind very well to a variety of different ion channels. The binding of the drug can be used first to identify a protein in its natural membrane and then in a further step to purify it. This procedure has been used successfully for a larger number of ion-channel forming proteins; a list of channels together with the drugs used for purification is given in Table 4.4. What we have described about the difficulties of functional purification also holds for affinity chromatography purification (at least partially), so again we will not focus on details of the purification procedures. A scheme for the purification of proteins by affinity chromatography is given in Figure 4.6. The best known example of an ion channel purified in this way is the nicotinic acetylcholine receptor (nAChR) protein from the electric organ of certain fish. We have already described a procedure to make a crude membrane fraction from this material (Section 4.2.1) which was optimized for obtaining chloride channels for reconstitution. A similar procedure can be used to prepare membranes for further purification of nAChR. The advantage of the starting material, membranes from electrocytes, is the high density of nAChR in these membranes (up to 50% of total protein). As shown in Table 4.4 α-bungarotoxin binds extremely well to acetylcholine receptors. In Figure 4.7 a procedure is outlined using α-bungarotoxin affinity chromatography to purify the nAChR from *Torpedo californica* electrocytes. This procedure includes the following steps:

1. isolation of the electric organ of *Torpedo californica*
2. isolation of electrocyte membranes
3. solubilization of total protein fom the membranes by detergents
4. binding of the nAChR to an α-bungarotoxin column

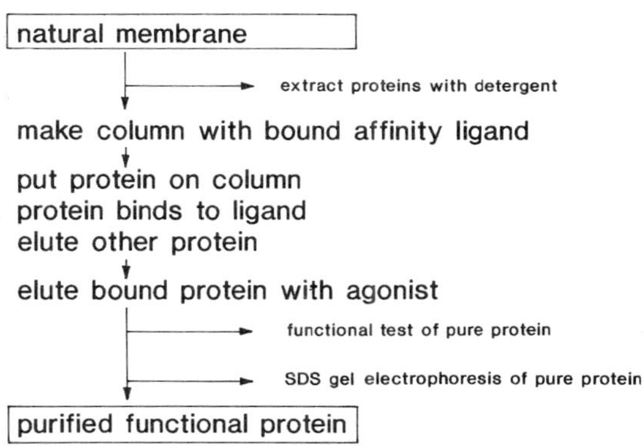


**Figure 4.6** Scheme of an affinity chromatography purification procedure applied to an unspecified protein. The specific binding of a drug to the protein is used as tool to identify and purify the protein.

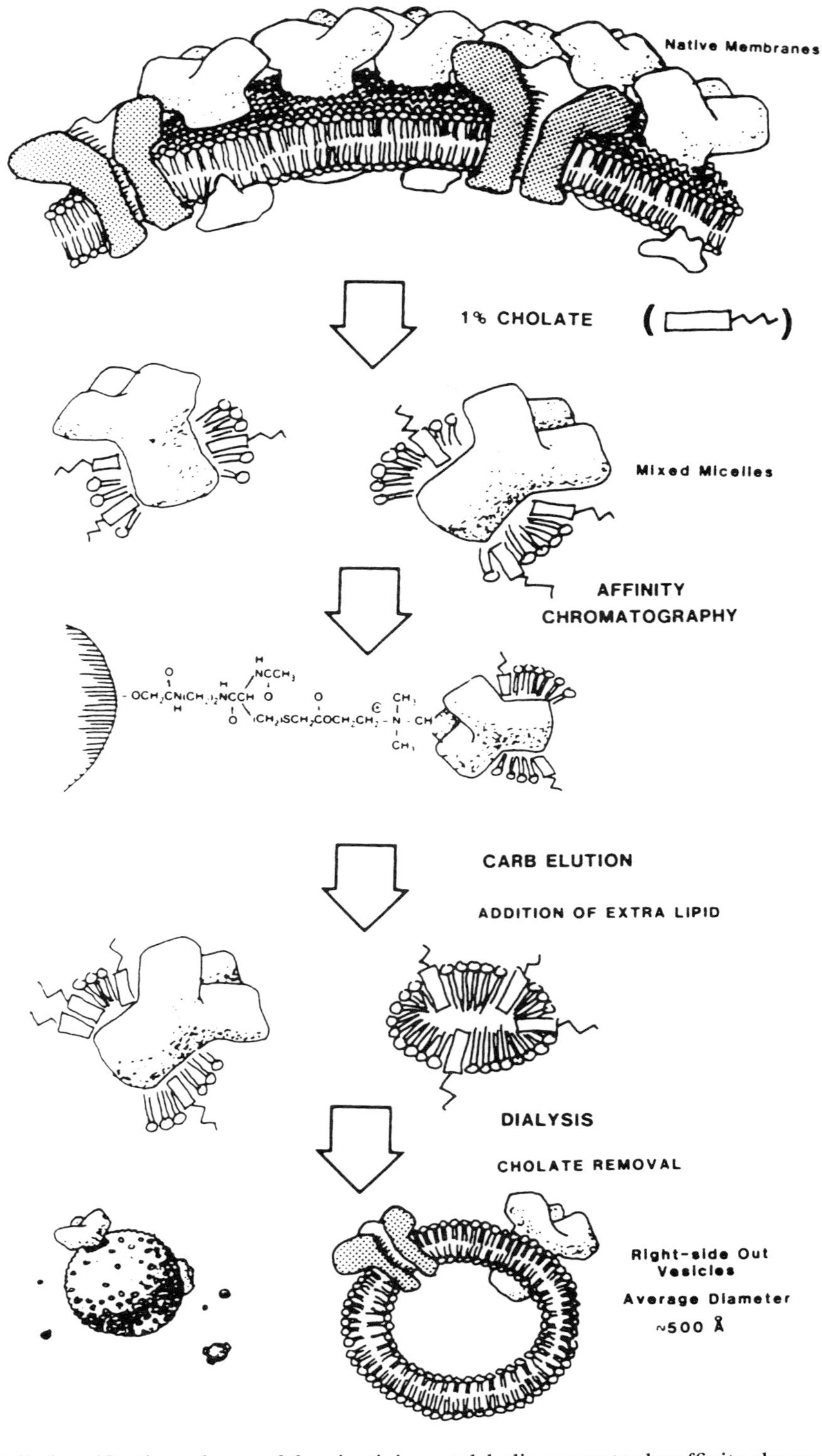


**Figure 4.7** Detailed purification scheme of the nicotinic acetylcholine receptor by affinity chromatography using α-bungarotoxin as ligand (Reproduced from McNamee *et al.*, 1986 with permission).

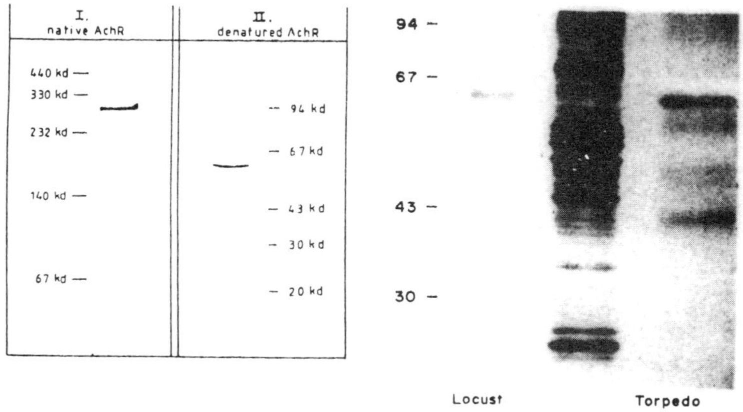


**Figure 4.8** SDS gels of different nicotinic acetylcholine receptors. The peripheral AChR and especially the AChR from electrocytes is composed of four subunits (α,β,γ,δ) as shown, the α-subunit is double in the intact protein. A nicotinic AChR from insect neurons (*Locusta migratoria*) shown for comparison is most probably only composed of one subunit (the intact protein is a homo oligomeric structure of four to five identical subunits).

5. elution of all the other proteins
6. elution of the nAChR from the column by adding a high concentration of agonist (for example, carbamylcholine 100 mM)
7. removal of detergents and incorporation of the protein into liposomal membranes
8. check of the final material on SDS-gel electrophoresis.

The purification of nAChR from electrocytes is standardized very well and ends with a protein composed of four subunits, as shown in Figure 4.8 on a SDS-gel. As a matter of fact the peripheral nAChR still is the best characterized ion channel and may serve as a textbook example of reconstitution of purified proteins (ion channels) into planar lipid bilayers. In Section 8.3, we will describe results of the reconstitution of nAChRs in more detail (Sobel *et al.*, 1977; Anholt *et al.*, 1981; Venter and Harrison, 1984).

### 4.2.3 Electroelution of polypeptides

As seen in the previous section, ion-channel forming proteins may be composed from one polypeptide (eventually forming a homo oligomeric structure), or from a number of different polypeptides, as for example the vertebrate muscle nAChR. A key question is what are the relations between polypeptide composition and channel function. The bilayer reconstitution technique here opens an interesting technical approach for additional experiments (Hanke *et al.*, 1990). The polypeptides as visualized on

| Native membrane |
|---|
| Purification of acetylcholine receptor |
| SDS-gel |
| Electroelution of polypeptides |
| Removal of SDS |
| Renaturation and assembly of protein |
| Binding studies |
| Electrophysiology (bilayer) |

**Figure 4.9** Scheme of electroeluting protein subunits (polypeptides) from denaturing SDS gels.

SDS-gel electrophoresis can be isolated and then can be reconstituted into bilayers in known stoichiometry. The principles of a procedure starting with SDS-gel electrophoresis to isolate polypeptides are given in Figure 4.9. In the following, some steps as used to isolate one subunit of a (most probably homo oligomeric) nicotinic acetylcholine receptor from insect neurons are listed:

1. Make preparation of purified AChR (for example from insect neurons, as this nAChR is a homo oligomeric protein).
2. Run mercaptoethanol SDS gel with purified material.

3. Stain bands on gel with $CuSO_4$.
4. Cut out the AChR-subunit band (here at about 65 kD) and remove stain.
5. Put material in dialysis bag and remove protein from gel material by electroelution.
6. Incubate electroeluted material with TSK-Fractogel DEAE and wash with buffer containing lipid and detergents which can be dialysed away.
7. Spin down and decant supernatant.
8. Repeat points 6 and 7 three times.
9. Remove protein from Fractogel with high salt solution (i.e. additional 0.8 M KCl).
10. Dialyse detergents against lipid as described elsewhere.
11. Let protein renature by keeping it for 3 days at 4°C.

In principle the procedure can be applied to any SDS-gel. A general limitation of the procedure is that steps of protein denaturation and renaturation are included, and it has to be verified that the protein's function is not changed by the use of these procedures. To date this has limited the technique to a few proteins; however, general applicability of the technique has been shown (Hanke *et al.*, 1990).

### 4.2.4 Incorporation of proteins into the membranes of liposomes

Very often, at the end of a protein purification, the protein has to be incorporated into a liposomal membrane. Furthermore, we will see later that such liposomes are a good starting point for reconstitution of proteins into planar lipid bilayers and for patch-clamp experiments. Thus we will give some hints here as to how to incorporate a protein purified using a detergent solution into a liposome membrane.

To transfer proteins from a detergent solution into a lipid membrane, two main tasks must be completed:

1. the detergent must be removed
2. the protein must be incorporated into a membrane.

A widely used technique for this is the dialysis of lipid against detergent. The protein preparation (with detergent) is put into a dialysis bag together with a certain amount of lipid. The dialysis bag is in principle a semipermeable membrane that excludes the passage of particles bigger than a selected size. This bag is immersed in a large volume of solution lacking detergent. Thus when the size of the detergent molecules is smaller than the pore size of the dialysis membrane it will be diluted into the detergentless solution due to the concentration gradient. When the dialysis membrane has been properly chosen, it will be impermeable to lipid and protein, thus they will stay in the bag. After a certain dialysis time the concentration of detergent in the bag will fall below a certain limit and the protein and the lipid will no longer be soluble in an aqueous solution. Therefore, for energetic reasons they will start to form liposomes. When the dialysis container has a big volume compared to the dialysis bag and when the aqueous solution is changed frequently, the detergent concentration can be diluted down close to zero. The vesicles remaining can be used for further experiments.

Other techniques to remove detergent are:

1. *Gel filtration techniques*: the detergent is removed by passing a mixture of protein, lipid and detergent over a gel-exclusion column.
2. *Dilution techniques*: the protein and lipid dissolved in detergent solution is diluted in a detergent free solution, the concentration of detergent becomes sufficiently low for the lipid together with the protein to start to form vesicles.

More details on detergent removal techniques are to be found in the literature.

The most important question relating to the above described dialysis technique is the choice of an appropriate detergent for protein solubilization (i.e. Helenius *et al.*, 1979), as some detergents (SDS, Triton X-100) can not or can only poorly be dialysed away. Others may affect the function of the protein to be solubilized. Thus detailed studies of this are required when setting up a protein solubilization procedure.

The most useful parameter to describe the ability of a detergent to be removed from an aqueous solution (especially by dialysis) is its critical micellar concentration (CMC); a high CMC is recommended when possible. Thus before starting a protein purification the detergent must be chosen not only according to the needs of making the protein soluble but also with regard to the question of how it can be removed. If a detergent has to be used that can not be dialysed away, additional steps as described in Section 4.2.3. and above must be included in the removal procedure.

## 4.3 HOW TO MAKE LARGE LIPOSOMES FOR PATCH-CLAMP EXPERIMENTS

Very often biochemical procedures end with microsomes, a liposomal type of preparation. The size of these liposomes depends on the biochemical techniques used to make them, however, liposomes are usually small (diameters in the submicrometre range). For certain groups of experiments, bigger liposomes are required, especially for patch-clamp experiments. The minimum requirement is that it should be possible to visualize the liposomes under the microscope. Additionally, for some types of bilayer reconstitution experiments bigger liposomes are better. Thus we will here describe briefly some techniques to make bigger liposomes out of preparations of small liposomes (Tank and Miller, 1983; Criado and Keller, 1987).

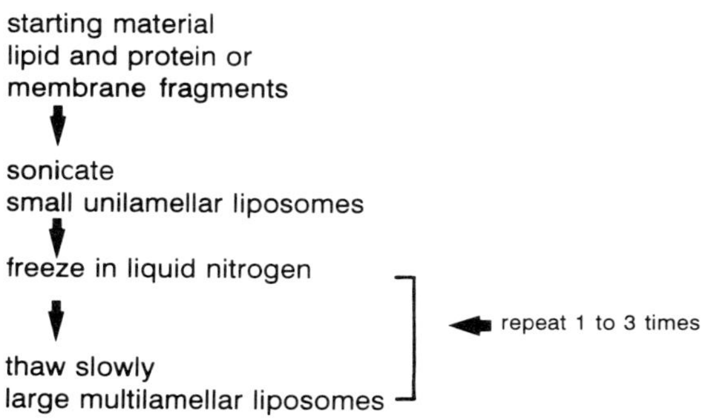


**Figure 4.10** By the freeze-thaw technique large vesicles can be made from small unilamellar liposomes. The resulting vesicles are large enough, up to 50 μm diameter, to be visualized under a microscope to perform patch-clamp experiments with them.

The most simple technique was first described by Kasahara and Hinkle (1977); they discovered that rapid freezing of small unilamellar vesicles (SUV) in liquid nitrogen and subsequent slow thawing induced the formation of large liposomes. The principles of this method are shown in Figure 4.10. These freeze–thaw liposomes are roughly spherical and can have diameters up to 50 μm. They can be used directly for patch-clamp experiments or for bilayer reconstitution experiments.

Another procedure for making large liposomes was introduced by Criado and Keller in 1987, the dehydration–rehydration technique. This technique includes the following main steps:

1. Dehydration of a small sample of concentrated small liposomes in a desiccator over $CaCl_2$; the liposomes are prevented from total drying by the presence of glycol in the aqueous solution.
2. Rehydration of the liposomes by addition of saline in a wet chamber overnight.

Large liposomes with diameters up to 100 μm are usually formed by this procedure within a few hours.

Finally, a third technique as introduced by Cohen and collegues in 1979 (Cohen *et al.*, 1979) is the formation of large liposomes on glass beads. A rough description of the technique is as follows:

1. Lipid dissolved in chloroform is dried in a round-bottomed flask using a rotary evaporator.
2. A defined amount of solution containing small liposomes with protein in their membranes is added.
3. 5–10 glass beads (1–3 mm diameter) are added and the flask is shaken for a few minutes.

This procedure will also result in large unilamellar vesicles containing the protein from the small vesicles incorporated into their membranes.

Of course all the above stated techniques can be combined and other procedures have been also described in the literature.

## 4.4 FLUX MEASUREMENTS WITH LIPOSOMES

Very often purified transport mediating proteins will be incorporated into the membranes of liposomes. Before doing bilayer reconstitution experiments with such preparations it can be very useful to test them for functionality of the proteins which will be later investigated in the bilayer. In principle, we have already used such a step in describing the functional purification of proteins (cGMP-activated channel) in the previous sections.

Of course nearly every method measuring the flux of a certain molecule or ion species out of or into liposomes will be useful. The following methods (besides others) have been described in the literature:

1. Optical techniques using dyes to trace some ions, especially calcium (for example Arsenazo III; see above).
2. Radioactive flux measurements.
3. The stop-flow technique; here the flow of ions across liposome membranes is started by rapid mixing of a liposome solution with an agonist and stopped after a defined time, the flux of the ions (or other material) is monitored by different techniques (for example optically with dyes).

Especially for methods 1 and 3, an amazing number of technical variations have been developed. For method 1, the use of fluorescent drugs specific for one ion (especially $Ca^{2+}$) has introduced a number of new approaches, such as $Ca^{2+}$-microfluorimetry (Mason, 1993). A good overview about the more classic rapid kinetic techniques (method 3) is given by M. G. McNamee *et al.* (1986).

After having tested a liposome preparation for functionality of the incoporated proteins, the risk of using a functionally impaired preparation in a bilayer reconstitution experiments is reduced significantly; therefore such tests are highly recommended.

## REFERENCES


Anholt, R., Lindstrom, J. and Montal, M. (1981) Stabilization of acetylcholine receptor channels by lipids in cholate solution and during reconstitution in vesicles. *J. Biol. Chem.* **256**, 4377–4387.

Benz, R. (1985) Porins from bacterial and mitochondrial membranes. *CRC Crit. Rev. Biochem.* **19**, 145–190.

Bergelson, L.D. (1980) *Lipid biochemical preparations.* Elsevier, Amsterdam.

Cohen, F.S., Zimmerberg, J. and Finkelstein, A. (1979) Fusion of phospholipid vesicles with planar phospholipid bilayer membranes. *J. Gen. Physiol.* **75**, 251–270.

Cook, N.J., Zeilinger, C., Koch, K.W. and Kaupp, U.B. (1986) Solubilization and functional reconstitution of the cGMP-dependent cation channel from bovine rod outer segments. *J. Biol. Chem.* **26**, 17033–17039.

Cook, N.J., Hanke, W. and Kaupp, U.B. (1987) Identification, purification and functional reconstitution of the cyclic GMP-dependent channel from rod photoreceptors. *Proc. Natl. Acad. Sci. USA* **84**, 585–589.

Criado, M. and Keller, B.U. (1987) A membrane fusion startegy for single channel recording of membranes usually non-accessible to patch-clamp pipette electrodes. *FEBS Lett.* **224**, 172–176.

Findlay, J.B.C., and Evans, W.H. (1987) *Biological membranes: A practical approach.* IRL Press, Oxford.

Hanke, W. and Miller, C. (1983) Single chloride channels from *Torpedo* electroplax: Activation by protons. *J. Gen. Physiol.* **82**, 25–45.

Hanke, W., Andree, J. Strohtmann, and Kahle, C. (1990) Functional renaturation of receptor polypeptides eluted from SDS polyacrylamide gels. *Eur. Biophys. J.* **18**, 129–134.

Helenius, A., McCaslin, D.R., Fries, E. and Tanford, C. (1979) Properties of detergents. *Methods Enzymol.* **56**, 734–749.

Jain, M.K. and Wagner, R.C. (1980) *Introduction to biological membranes.* John Wiley, New York.

Kasahara, M. and Hinkle, P.C. (1977) Reconstitution and purification of the D-glucose transporter from human erythrocyte. *J. Biol. Chem.* **252**, 7384–7390.

Kates, M. (1986) *Techniques of lipidology.* Elsevier, Amsterdam.

Labarca, P., Coronado, R. and Miller, C. (1980) Thermodynamic and kinetic studies of the gating behaviour of a K-selective channel from the sarcoplasmic reticulum membrane. *J. Gen. Physiol.* **76**, 397–424.

Latorre, R. (1986) *Ionic channels in cells and model membranes.* Plenum Press, New York.

Mason, W.T. (1993) *Fluorescent and Luminescent Probes for Biological Activity*. Academic Press, London.

Miller, C. (1986) *Ion channel reconstitution*. Plenum Press, New York.

Sakmann, B. and Neher, E. (1983) *Single-channel recording*. Plenum Press, New York.

Schlindler, H. and Quast, U. (1980) Functional acetycholine receptors from *Torpedo marmorata* in planar membranes. *Proc. Natl. Acad. Sci., USA* **77**, 3052–3056.

Sobel, A., Weber, M. and Changeux, J.P. (1977) Large-scale purification of the acetylcholine receptor in its membrane-bound and detergent-extracted form from *Torpedo marmorata* electric organ. *Eur. J. Biochem.* **80**, 215–224.

Stahl, E. (1969) *Thin layer chromatography*. Springer, New York.

Tank, D.W. and Miller, C. (1983) Patch-clamped liposomes: Recording reconstituted ion channels. In: *Single channel recording*. B. Sakmann and E. Neher, eds. Plenum Press, New York, pp. 91–106.

Tank, D.W., Miller, C. and Webb, W.W. (1982) Isolated patch-recording from liposomes containing functionally reconstituted chloride channels from *Torpedo electroplax*. *Proc. Natl. Acad. Sci. USA* **79**, 7749–7753.

Tareilus, E., Hanke, W. and Breer, H. (1990) Identification of a cationic channel in synaptosomal membranes. *Eur. Biophys. J.* **19**, 217–236.

Venter, C. and Harrison, L. (1984) *Receptor biochemistry*, Vol. 2, *Receptor purification studies*. Alan R. Liss, New York.

CHAPTER FIVE

# Methods for making planar lipid bilayers

Having outlined the theoretical basis of planar lipid bilayers and membranes and the technical prerequisites of bilayer experiments, in Chapter 4 some ideas about the biochemical requirements of bilayer experiments were given. In these preceding chapters we have established the fundamentals for proceeding to the main question of this volume: how to make planar lipid bilayers and how to perform experiments with them. A variety of techniques has been developed to make planar lipid bilayers (Ti Tien, 1974; Hanke, 1985). We will discuss the more important of them together with the special technical needs of each. Additionally, for some of the techniques we will give simple instructions on how to make the bilayer. Mainly we will follow the historical development of bilayer research when discussing the different techniques that have been described in the literature. To give a complete overview of bilayer techniques, we have included a section about patch-clamping large liposomes, a technique which is placed somewhere between bilayer and patch-clamp experiments. As this volume is addressed to bilayer techniques, however, not many details about patch-clamping are given, these being readily available elsewhere (Hamill *et al.*, 1981; Sakmann and Neher, 1983).

In this chapter we want to cover the standard bilayer techniques and procedures necessary for those without experience to get started with bilayer experiments. For some more sophisticated experiments, and proceeding to the frontiers of bilayer research, technical details of advanced bilayer techniques and some recent developments in the field are given in Chapter 8.

## 5.1 PAINTED LIPID BILAYERS

### 5.1.1 Use of optics

The core concept in a planar lipid bilayer experiment is to span an aperture of an area of let us say 1 $mm^2$ with a double layer of pure lipids. To stay in line with the ideas given in the introduction, it is of course best to use one pure, synthetic lipid to do this. As lipids are not soluble in water, they are usually handled dissolved in organic solvents. This may have been one of the starting points of Mueller and coworkers (1962) when they did the first bilayer experiments in the early 1960s. They started with a double compartment chamber (see Section 3.1.2.); the two compartments of the chamber being connected by an aperture in the wall between them and each being filled with aqueous solutions. Then they spread a drop of lipid dissolved in an organic *non-polar* solvent across the hole and thinned it out, using a mechanical tool. By this method, the layer of lipid/solvent can be made so thin that its thickness is in the range of the wavelength of visible light, which is confirmed by observing the membranes with a small telescope.

Figure 5.1 shows a sketch of such an optical set-up. The membrane is illuminated diagonally and is observed in the reflected light. When the light passes through the lipid layer it has to pass two optical interfaces, from water to the organic phase and from the organic phase to water. At such interfaces between areas of different optical density, light is reflected. At the lipid layer there are two reflections, one without and one with a phase jump of 180°. The reflected light from the lipid layer is a sum of both reflections. The superimposition of the light from both reflections may give partial wipe-out of the light (when white light is used), if the thickness of the membrane is within the range of the wavelength of the light (comparable effects happen with soap bubbles). In the experiments, white light is used and this type of membranes gives nice rainbow colours during observation. (Inverting the arguments, simply by the fact that these rainbow colours are observed, it is obvious that the thickness of such membranes must be in the range of the wavelength of visible light.)

When observing the coloured membranes for some time, usually minutes, black spots occur spontaneously on them, usually at the lower margin of the film. Figure 5.2 shows a sketch of a coloured membrane without black spots, some minutes later with the spots and finally in the totally black state. The explanation for this optical observation is that organic solvent is diffusing out of the membrane. After most of the solvent has left the membrane, only a lipid bilayer, or even better a system with a small amount of solvent, remains. The membrane is now very thin due to the small amount of remaining organic solvent in the system; it is a lipid bilayer with some microlenses of organic solvent. (We will discuss the forces which are responsible for the thinning of the bilayer later in this chapter.) The thickness of a true lipid

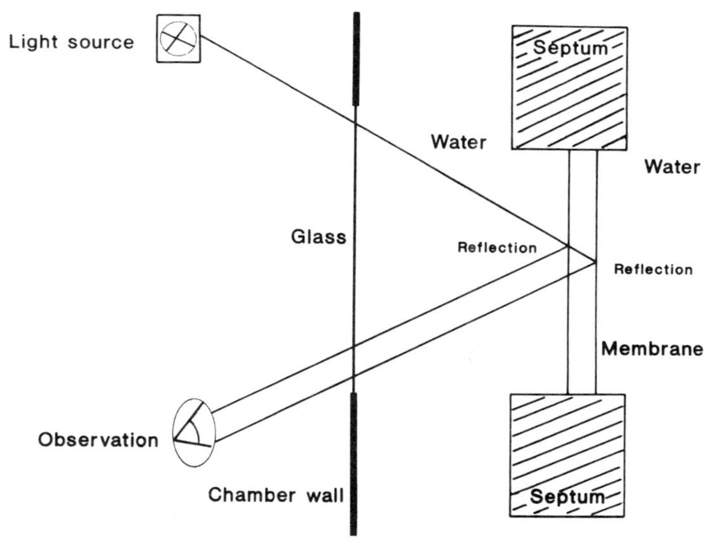


**Figure 5.1** Optical observations of the formation of a painted bilayer from a coloured film. A painted bilayer is illuminated diagonally and observed in reflected light. Two reflections, one with and another without 180° phase-shift, occur. The sum of the two reflected beams from white light gives rainbow colours with thick films (wavelength of light) and total wipe-out (black lipid bilayer) with real bilayers (which are 3–5 nm thick).

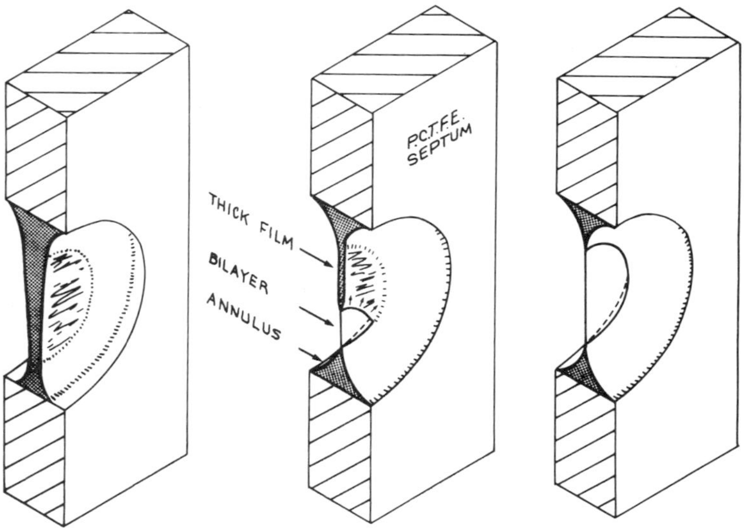


**Figure 5.2** A sketch of a black lipid bilayer and its formation from a thin lipid film. First, the film contains a lot of solvent (left). Part of the solvent diffuses out of the membrane and a black spot appears at the lower margin of the film (centre). Finally when most of the solvent has left the film, the membrane is black all over (right).

bilayer, from theoretical calculations, should be about 3–5 nm, depending mainly on the length of the fatty acids or alcohols of the lipids used. Compared to the wavelength of visible light used for observation (400–700 nm) this gives double reflection at nearly the same point, once without and once with phase jump, and thus total wipe-out. In reality this is found in the black spots, due to the fact that here the thickness of the membrane is small compared to the wavelength of visible light. After a longer time the entire membrane becomes a true bilayer and therefore black all over (some ten minutes). For this reason when these bilayers were first produced they were called 'black lipid membranes'.

Figure 5.3 shows how to make painted planar lipid bilayers from lipid dissolved in decane. Although the principle is very simple, a drop of lipid dissolved in organic solvent (decane) being spread over a small hole separating to aqueous phases, many details have to be taken into account, which will be discussed in the following.

### 5.1.2 Choice of lipid

For the beginner, starting the first bilayer experiments, the question of the choice of the lipid to be used (see Chapter 4) is not that important because the principles of bilayer formation are not (or at least are only weakly) dependent on the lipid composition (some details will be discussed in Section 8.1). A cheap, easy-to-handle lipid which is commercially available at low price is partially purified soya bean lecithin (Asolectin, see Chapter 4), which is a good choice not only for starting with the experiments but is widely used in reconstitution experiments too. Other lipids, depending on the needs of special experiments, may of course be used as well. In most bilayer experiments, phospholipids

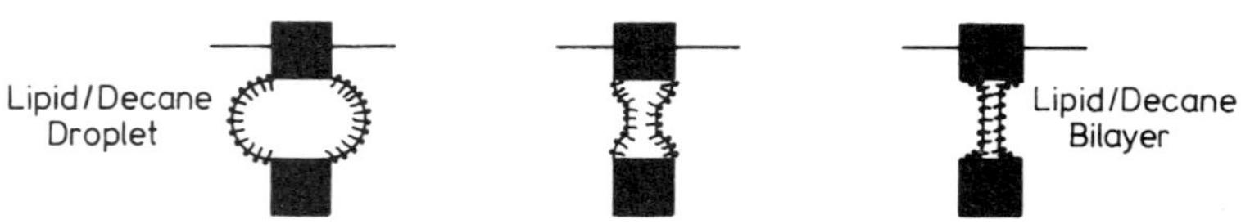


**Figure 5.3** The formation of a painted bilayer on a partition. A drop of lipid is spread on the aperture in the partition. By a number of different forces, the lipid diffuses out of the membrane which thus finally becomes a black lipid bilayer.

are used, as they are easy to handle and are the main lipid components of biological membranes. Furthermore, they can be synthesized without problems and are commercially available at acceptable price and in high quality (see Appendix 4).

### 5.1.3 Organic solvents

The other main component in such an artificial membrane is the organic solvent used to dissolve the lipid. Mainly, decane is used as the non polar solvent in painted bilayer experiments, but other solvents may be necessary and can be used for special experiments (long-chain alkanes, squalene). It is not appropriate to use highly volatile solvents as they are difficult to handle in painted bilayer experiments. Thus, for example, hexane or pentane are rarely used with painted bilayers, but nearly exclusively with folded bilayers.

Furthermore, the solubility of organic solvents in lipid has to be taken into account. When measuring the electrical properties of a painted bilayer made from a decane solution, it was found that the specific capacity is about 0.4 $\mu F/cm^2$, which is much less than expected from theoretical calculations for a true bilayer (about 0.8 $\mu F/cm^2$). The reason is that a significant amount of decane remains dissolved in the bilayer in form of microlenses (as stated above). Other solvents are less soluble in the lipid and thus give somewhat higher specific capacities (e.g. hexadecane, see later). Usually the specific capacity of painted bilayers is higher when the length of the carbon chain of the organic solvent is longer; obviously less solvent remains in the membrane.

However, the use of organic solvents with long chains is limited by practical reasons as they have their phase transition at comparably high temperatures, so they may be solid at room temperature. On the other hand, some methods use this knowledge to get rid of another part of the solvent in a planar lipid bilayer. Hexadecane freezes out of the bilayer when cooled below its melting point (18°C). The amount of solvent remaining in the membrane is dependent on the time the bilayer is observed after formation. The effect of this thinning out of the bilayer has to be taken into account when performing bilayer experiments on a longer time-scale. When virtually solvent-free bilayers are needed it seems to be not very useful to take advantage of this effect; it is better to use another technique to make the bilayers (discussed later in this chapter).

As the stability of bilayers is typically in the range of minutes to hours, to wait for an hour or so to thin out the bilayer before starting an experiment, may end the experiment before it has really begun. Additionally, it has been found that the use of some solvents which are less soluble in the lipid often results in more unstable bilayers and difficulties in handling; thus decane has been demonstrated to be the best choice for most experiments with painted bilayers and we will focus on this solvent here.

The above discussion is only relevant, when the amount of organic solvent in a bilayer is relevant for the experiment. Nevertheless one should have in mind two things: first, coming back to our arguments from the introduction about one-component systems, addition of any solvent represents the addition of another component to the system; and second, a biological membrane, which we are in principle interested in and want to simulate, by definition does not contain organic solvents. For these reasons, it would in principle be much better to make solvent-free bilayers, something we will describe later.

### 5.1.4 Apertures

So far we have neglected the properties of the hole which has to be made for the bilayer in the partition between the chambers. As described in Chapter 3, for painted bilayers plastic or Teflon chambers are widely used and an aperture with a diameter of 0.1 to 1 mm or even larger is drilled into the partition, which is about 1 mm or less thick. The mechanical quality of these apertures is very important for the stability of the bilayers, and any ripple of the boundary of the aperture, even in the far sub-millimetre range is big compared to the thickness of a bilayer. As the quality of the boundary of the aperture is limited by technical reasons to a ripple in, let us say in the micrometre range, bilayers

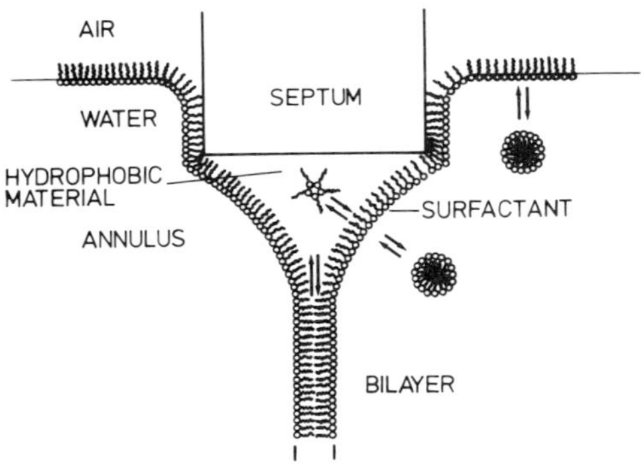


**Figure 5.4** Structure of a bilayer at its contact with a thick partition. As the septum is very thick compared with the bilayer itself, an annulus of organic material is necessary to compensate for the difference in thickness. In addition, lipid monolayers cover the hydrophobic material of the partition (and may be found at the air–water interface). The material of the annulus and the lipid are in equilibrium with micelles in the water phase. (Reproduced and modified from White, 1986, with permission)

are usually not stable on these holes. Second, even the thickness of a thin partition is orders of magnitudes higher than the thickness of a real bilayer. There has to be an annulus of *non-polar* solvent or lipid to compensate for this difference, as shown in Figure 5.4. It has been found that a proper pre-treatment of the hole with lipid significantly increases the stability of the bilayers. In most experiments with painted bilayers the pre-treatment is done as follows:

1. A drop of lipid dissolved in a volatile solvent, usually decane to avoid an additional organic component in the system, is added to the hole, covering it (the solution used for the bilayers itself can be used).
2. The solvent is evaporated by a stream of nitrogen (to avoid denaturation of the lipid).
3. Instead of lipid, squalene or vaseline, for example, may be used, but it should be borne in mind that another component is added to the system by doing this.

After pre-treatment, the chambers can be used as described to add a drop of lipid to the aperture. When a bilayer breaks during an experiment, very often it is not necessary to add new lipid. Simply re-spreading the lipid remaining on the partition close to the hole with a mechanical tool may be sufficient to make a new coloured film.

### 5.1.5 Example preparation of a painted bilayer

A more complete simple recipe to make a painted planar lipid bilayer is given here.

1. First dissolve 30 mg of lipid in 1 ml of decane; at the beginning it may be adequate to use a lipid extract from soya bean membranes (Asolectine) because this is easy to handle and cheap; for better defined systems a synthetic lipid may be fine.
2. Store the lipid on water and ice during the experiment to avoid denaturation.
3. Pre-treat the aperture in the chamber as described above.
4. Fill the chambers with aqueous solution (for example 1 M KCl or standard Ringers solution) so that the aperture between the two chambers is completely covered.
5. Add a drop of lipid solution to a Teflon coated metal loop (or any other tool used to spread the lipid on the aperture), using a pipette.
6. Carefully bring this drop through the water surface to the aperture.
7. Spread the drop of lipid solution on the aperture under optical control so that rainbow colours can be seen, when observing it in reflected light. This is necessary because the bilayer will not form spontaneously on the aperture; the lipid must be spread.
8. Wait for the occurrence of black spots; it has been found by some workers that bubbling the coloured film smoothly with small air bubbles using a glass pipette is an easy way to thin out the film faster by spontaneous thinning.
9. If the electronics are already installed, measure the capacity of the membrane at different states and monitor the conductivity of the system.
10. When the membrane has become black all over (when controlling the membrane capacity it should be greater than 0.4 $\mu F/\mu m^2$ due to a possible curvature of the bilayer) a bilayer has been made.

11. In a later step, the capacity of the bilayer can be minimized by changing the height of aqueous solution on one side of the bilayer to reduce a possible curvature of the bilayer.

Point 5 may be modified; we have used with good success a small coated metal loop as described, but other tools may be as helpful (for example small brushes have been widely used).

The above procedure should be practised again and again, as there is a high probability that the membrane will break frequently and so give an insight into the frustrations of bilayer experiments. Here it should be stated again that one should extremely carefully clean all parts used in the experiments, as this significantly affects the stability of bilayer experiments. Due to the importance of clean equipment, especially chambers and partitions, one should remember the cleaning procedures given in Chapter 3.

Besides the equipment described in the methods above and in Chapter 3, many other set-ups have been used to perform painted bilayer experiments. First, the geometry of the set-up may be changed over a wide range, depending on the needs of the actual experiment. Chambers have been used from 0.1 ml volume up to some 10 ml volume. The diameter of the aperture on which the bilayer is made, may also vary from 0.1 mm to 3 mm. Spherical painted bilayers, for example, can be made, by blowing a bubble from a drop of lipid solution on the end of a thin plastic or glass tube immersed in aqueous solution (Ti Tien, 1974; Lühring, personal communication). As we are not able to present an encyclopaedic treatment of all details of bilayer techniques here, but just want to present some more general approaches, the reader is referred to the literature for further details (van den Berg, 1965; Jain, 1972; Finkelstein, 1974; Ti Tien, 1974; Fettiplace *et al.*, 1975; Hanke, 1985; Miller, 1986).

### 5.1.6 Theoretical aspects

Before proceeding to the folded bilayer techniques, we want to discuss a few theoretical aspects of painted planar lipid bilayers and their consequences for mechanical handling and stability. This discussion is about painted bilayers but also holds true, at least partially, for all other bilayer types. As can be seen from Figures 5.2 and 5.4 a bilayer system is made from a number of different parts and components, for example the bilayer itself, which may contain microlenses of solvent, and the chamber with the partition. Thus different boundaries, between water and lipid, annulus and lipid, and lipid and wall, are given. An annulus created of some hydrophobic material, which is necessary for stability, has to be taken into account to compensate for the differences in thickness of film and partition.

When making a bilayer on such a hole, the area of the hole is

$$A = r^2\pi \qquad [5.1]$$

where $A$ is the area and $r$ the diameter of the hole. The real area of the bilayer must be calculated, for example, for the determination of the specific capacity, which is

$$C_s = C/A \qquad [5.2]$$

where $C_s$ is the specific capacity and $C$ is the measured capacity of the bilayer. This area is not necessarily identical to $A$, as the area of the annulus has to be taken into account and as the bilayer may have some curvature. Theoretically the measured capacity of a bilayer is to a first approximation that of a flat capacitor with

$$C = \varepsilon_o \times \varepsilon_r A/d \qquad [5.3]$$

where $d$ is the thickness of the film, $\varepsilon_o$ is the absolute dielectricity constant and $\varepsilon_r$ is the relative dielectrical constant of the medium between the two plates of the capacitor, which in this case are the water phases. Here the medium is the hydrophobic core of the bilayer together with the solvent, usually fatty acids from the lipid and decane, and for an approximation $\varepsilon_r$ is taken to be close to 2–3 (for water the value is about 80). The only way to get an acceptable estimate of the bilayer area is to measure it with adequate optics, after having levelled out the aqueous solution levels very carefully to minimize surface curvature of the bilayer. This can best be done by measuring the capacity and varying the level in one chamber until a minimum capacity is reached (see Section 5.1.5).

For the stability of a membrane, coloured or

black, the forces which are acting on it are important. For example, it is necessary to know which forces provide for the thinning of a coloured film to a black lipid membrane. The lipid of the system in Figures 5.2 and 5.4 lowers the tension at the interface between bulk solution of the annulus and the aqueous solution according to

$$d\tau = -\Gamma\, d\mu \qquad [5.4]$$

with

$$\mu = \mu_o + RT \ln a \qquad [5.5]$$

where $\mu$ is the chemical potential of the lipid with its activity $a$ in the annulus, $\Gamma$ is the concentration at the interface (mol/cm$^2$), $\mu_o$ is the reference chemical potential, $T$ is the absolute temperature and $R$ is the gas constant. A coloured film spread from a bulk solution with a proper tool, thins spontaneously by the action of some driving forces. The main one at the beginning is the Plateau-Gibbs border suction:

$$\Delta P = 2/\tau/r_1 \qquad [5.6]$$

where $r_1$ is the radius of the curvature of the annulus. By this force, solvent from the membrane flows to the annulus. When the remaining thickness of the membrane is in the range of a few hundred ångströms the London–van der Waals attraction between the two aqueous phases separated by the film becomes important. It is

$$\Delta F = -H/12\,\pi d^2 \qquad [5.7]$$

where $d$ is the thickness of the film and $H$ is the Hamaker coefficient. This force usually is responsible for the final thinning of the film.

All the forces on the planar lipid bilayer have to be in an equilibrium to reach a stable state of the system; usually this is not the case, especially as many additional forces which are present in the system can not be controlled adequately. Nevertheless, bilayers may be very stable systems in a metastable state, which may last for hours. In principle the above described facts about the forces acting on a bilayer should be sufficient for successful experiments; readers more interested in the theory should screen the literature (Ti Tien, 1974; White, 1986).

The measurement of the capacity of a bilayer is, besides the measurement of its conductivity, the most important tool for controlling the system. The capacity of the bilayer device and other effects have to be taken into account when doing this, but if one measures first the capacity without bilayer (using a drop of solvent covering the aperture to close it) and then with bilayer and takes the difference, an acceptable value is usually obtained. Some details of electronic devices to measure bilayer capacities have been described in Chapter 3.

The current flowing across a bilayer at a fixed potential is the other important value giving information about the quality of the bilayer formed. If fluctuations are present in a pure lipid bilayer at comparable low voltage (less than ±75 mV), this indicates either impurities from other components in the lipid or instability of the bilayer, which may be due to boundary leaks, for example. Although the current measurement is not good evidence for real bilayers, drops of lipid having about the same resistance as bilayers, it is necessary to control the bilayer quality by it. In real experiments, bilayers often have good capacities and optically look black, but are 'noisy' and unstable and so are of no use. In such a case, it is best to break the bilayer and start a new experiment.

## 5.2 FOLDED BILAYERS

### 5.2.1 Folded bilayers from lipid solutions

As stated earlier, one disadvantage of painted bilayers is the amount of solvent remaining in the bilayer. For this reason Montal and Mueller (1972) developed a technique to partially avoid the problem of organic solvents. Their idea was that a planar lipid bilayer is technically the sum of two lipid monolayers. Lipid monolayers can easily be made from lipid solutions on air–water interfaces and have been investigated in depth, for example on film balances as described in the literature (e.g. Adamson, 1976; Duschl *et al.*, 1989). In fact the concept of such experiments had been published long before the first real use of the technique in the 1960s by Takagi and

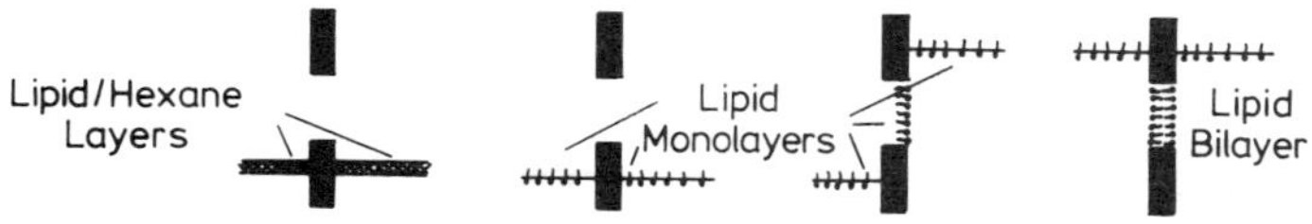


**Figure 5.5** Formation of folded bilayers from two monolayers, spread from lipid–hexane solutions at air–water interfaces. As shown on the the left side, one starts with spreading the monolayers below the aperture in a very thin partition. Then, after evaporation of the solvent, the bilayer is formed by raising the aqueous levels to above the aperture (simultaneously or one after the other).

coworkers (1965), who outlined many principles used later in folded bilayer experiments.

In general it should be possible to make two monolayers on two sides of an aperture separating two aqueous phases and to combine them to a bilayer over the aperture as shown in Figure 5.5, resembling the function of a zipper. The monolayers are made below the hole and then the aqueous levels are simply raised over the hole. It has been demonstrated that this is possible when a few technical details are taken into account (Montal and Mueller, 1972). The monolayers should be spread from a highly volatile solvent to reduce the amount of solvent remaining in the lipid phase; hexane or pentane have been found to be a good choice. A sufficient time should elapse after spreading the lipid, to give time for the solvent to evaporate, 5–15 min are usually sufficient. After that the levels of the aqueous solutions should be raised very carefully to avoid mechanical distortion. Whether the surfaces are raised at the same time or consecutively seems not to be significant. Usually a syringe is used to change the levels of aqueous solution in such experiments. When such a bilayer breaks, in most cases it is sufficient to lower the level on one side and raise it again to get a new bilayer, as when lowering the level, the first monolayer is established over the aperture. This is due to the fact that because of surface forces the water does not flow through the very small aperture and an air–water interface is thus established at the aperture when the water level is lowered. Lipid remains on this interface when lowering the water level, thus forming the first monolayer.

Following this outline, bilayers with specific capacities of about 0.8 μm/cm$^2$ have been made, so there can not be much solvent remaining in the lipid. Nevertheless, a small amount of hexane or pentane will of course remain dissolved in the lipid, so the bilayers are not really solvent-free. Again the stability of the bilayers is dependent on the quality of the membrane carrying aperture. In experiments with folded bilayers, Teflon chambers are usually used as well as Teflon partitions, as other materials are not chemically stable in contact with hexane or pentane. The partitions used are usually very thin, 5–25 μm, and the apertures have diameters of 0.1–0.5 μm. The production of the septa (partitions) carrying the aperture has been described in Chapter 3 in detail, together with details about the chambers and other equipment.

As with painted bilayers a pretreatment of the apertures is necessary to compensate for boundary ripple and the difference in thickness of septum and membrane, but with folded bilayers a pre-treatment with lipid has not been found to be appropriate. Instead the apertures are pretreated with a solution of a less soluble organic component; hexadecane, silicone, vaseline or squalene have been used, for example, to add the non-polar material for the annulus. These compounds are dissolved in an organic solvent which is highly volatile, hexane and chloroform have been used here. A drop of the solution is spread on the partition and the solvent is evaporated with a stream of nitrogen or air, then the partitions can be used as described. Of course the material from the pre-treatment is, but not very much, soluble in the lipid of the bilayer, adding some solvent to it. For this reason and due to the remaining solvent after spreading the lipid, as discussed above, the folded bilayers are close to, but not entirely solvent-free, therefore they are often called 'virtually solvent free bilayers'.

In Chapter 3, which described the technical needs of bilayer experiments, the mechanical components of an experiment with folded bilayers were shown: the chambers, a clamp to hold the two chambers together, the partition with the aperture and a syringe to raise the levels of the aqueous solutions. A complete set-up was shown there, too. Following our outline from above, a procedure for an initial experiment follows:

1. Dissolve 2 mg of lipid (again Asolectin) in 1 ml of pentane or hexane.
2. Store the solution on water-ice.
3. Make a solution from vaseline (or silicon) in hexane (as much as can be dissolved).
4. Add a drop of the vaseline solution to the hole (aperture) in the septum using a pipette and evaporate the hexane under a stream of air or nitrogen.
5. Mount the chambers and partition as described in Chapter 3.
6. Fill the chambers with aqueous solution (1 M KCl or standard Ringer's solution) to a level just below the aperture in the partition.
7. Add approximately 20 µl of lipid solution per square centimetre of chamber surface to the air–water interfaces (both sides).
8. Wait 15 min to allow the organic solvent to evaporate.
9. Raise the aqueous levels using a syringe (carefully, to avoid mechanical distortion); for practical reasons, raise first the back and then the front level.
10. Monitor electrically the capacity and conductivity of the system.
11. In successful cases, after having raised the levels on both sides of the hole to above the hole, a bilayer should be established.
12. Minimize the bilayer capacity by changing the aqueous level on one side of the bilayer, to reduce bilayer curvature.

When at point 7 different lipids are added to the surfaces at both sides of the partition, this technique enables the experimenter to make asymmetric bilayers, which can not be done with painted bilayers. This was another idea of Montal and co-workers (1972) and Takagi *et al.* (1965) when they created this technique and gives it advantage over the classical technique of painted bilayers, which does not allow asymmetrical bilayers. Asymmetrical bilayers are of high interest as all biological membranes are asymmetrically structured and in principle we do want to investigate models of biological membranes.

Instead of raising the aqueous levels from below the aperture to above, some experimenters have used equipment to draw a Teflon partition with the aperture *into* an aqueous phase with a lipid monolayer on top of it. In principle this technique is identical to the technique described; however, the set-up is more difficult to handle and one must be careful to get a good electrical isolation between the two chambers (which usually is required). As stated in Section 5.1, the variation of set-up layouts described in the literature is remarkable and a complete overview is not possible.

Later we will discuss a more sophisticated set-up enabling the advanced experimenter not only to control the lipid composition of each side of the bilayer separately, but in addition the surface pressure of each of the monolayers the bilayer is made from (Section 8.1). In fact, from the beginning of the making of folded bilayers it has been one of the aims to create a technique that enables the scientist to create bilayers from controlled monolayers. The techniques are available now, but the set-up is highly complicated and for most experiments the simpler procedure described here is sufficient.

The remaining requirements of the experiments are just the same as for all bilayer experiments. Again with respect to the cleanness of the system one has to be very careful. Most of the theoretical basis of the membrane treatment is identical to that for painted bilayers.

### 5.2.2 Folded bilayers from vesicle suspensions

Obviously even folded bilayers are not really solvent-free and, as stated earlier, it may be necessary to minimize the amount of solvent in the bilayer. With the standard techniques of bilayer formation to date, there has been no way to avoid the pre-treatment of the aperture in folded bilayer experiments. It may be possible when using very small holes with diameters less than 20 μm and very thin partitions with a thickness in the micrometre range to avoid a pre-treatment, but these experiments have not been very successful and conditions at the annulus are not stable. So the only remaining way is to reduce the solvent in the lipid solutions the monolayers are made from. In the 1970s, Schindler and Feher (1976) and Schindler (1980) used the effect that stable monolayers are formed on the air–water interface of vesicle suspensions by themselves. The time course of the monolayer and the final lateral pressure to establish them depend on the composition and the properties of the vesicle solution. Details of experiments to investigate these monolayers on film balances have been described in the literature and were introduced by K. Kaufmann (Göttingen, FRG) (Kaufmann, 1989).

We will not discuss methods of making the liposomes here in detail, the interested reader may use the literature, and the procedures described in Chapter 4 supply some basic information. In outline the procedure to make bilayers from monolayers on top of vesicle suspensions is as follows.

1. Dissolve 3 mg of Asolectin in hexane (1 ml).
2. Evaporate hexane in a small flask under continous motion, thus forming a thin lipid film on the glass.
3. Add 1 ml of aqueous solution to the flask.
4. Sonicate the Asolectin in the aqueous solution with a Branson sonicator (30 pulses of 1 s with breaks of 1 s at high power).
5. Hold the solution on ice during sonication to avoid heating.
6. As it has been shown that bilayers are better formed from monolayers with comparable high lateral pressure, the small sonicated liposomes are made larger by a freeze–thaw procedure. This is helpful in the light of the data discussed in Chapter 4, as larger liposomes give monolayers with higher lateral pressure in shorter time.
7. The freeze–thawing is most easily done by freezing a sample of the liposome solution in liquid nitrogen and then thawing it on water–ice slowly.
8. The procedure may be repeated two or three times and results in large multilamellar vesicles, as can be seen partially from the opalescence of the liposome solution.

After having made a liposome solution, about 20 μl of this solution per millilitre of aqueous solution should be added to the bilayer chambers and then a sufficient time should be waited to establish a stable monolayer, about 30 min for the conditions given. The remaining step is to raise the surfaces and to establish a bilayer the same way as described for bilayers made from monolayers out of lipid–hexane solutions (see earlier). In Figure 5.6 a diagram of this technique is given. In other respects the way to establish the monolayers is identical to the Montal technique of folded bilayers, and the same equipment is used.

Besides producing bilayers with a very small amount of organic solvent (virtually solvent-free bilayers), the above described experiments of folded bilayers gave some first insights into the properties of the monolayers which are necessary to form stable bilayers. The main finding was that there is a lower limit for the lateral pressure of the monolayers in the range of 25–30 dyne/cm$^2$ to give stable bilayers. Controlling the surface pressure of the monolayers during the experiment would be of interest, but is not possible with the technique described here. On the other hand, it has been shown that surface pressure is extremely sensitive to the composition of the bulk solution (and to other chemical and physical parameters). Therefore the spreading

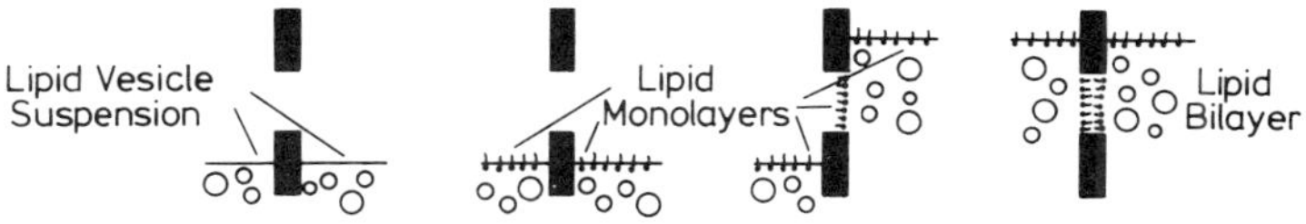


**Figure 5.6** Formation of folded bilayers from monolayers spread from liposomes. Apart from the technique used to make the monolayers, this method of bilayer formation is identical to that shown in Figure 5.5.

properties of a system have to be investigated prior to doing bilayer experiments with a film balance. Alternatively, data must be available from other experiments or the literature. Best, however, would be to measure the surface pressure of the monolayers during the bilayer experiment. We will come back to this point later.

## 5.3 DIP-STICK BILAYERS

Using the techniques described, a large number of experiments to characterize and to use planar lipid bilayers have been done during the last decades. At the end of the 1970s, however, another new electrophysiological technique strongly affected biological research, the patch-clamp technique (see Introduction; Hamill *et al.*, 1981, Sakmann and Neher, 1983). The principle of this technique is to electrically isolate a small area of a membrane under the tip of a glass pipette electrically and then to measure its properties. This has been done successfully with a variety of biological membranes, by patching areas in the range of square micrometres. The principles of the patch-clamp technique are given again in Figure 5.7 (see also Figure 1.7). For detailed information the interested reader is referred to the original literature.

The seal between a biological membrane with a clean surface and a glass pipette may be good enough to result in giga-ohm electrical resistance. Some main advantages of electrically isolating a very small membrane area under a glass pipette are the resulting low noise in electrical measurements and the fact that no additional components are needed to establish the giga-ohm seal (a pre-treatment of the pipette tip adding another component to the system is not necessary). It is tempting to try to apply this technique to pure lipid bilayers to exploit all its advantages. Technical procedures have been developed by different groups during the last few years to do

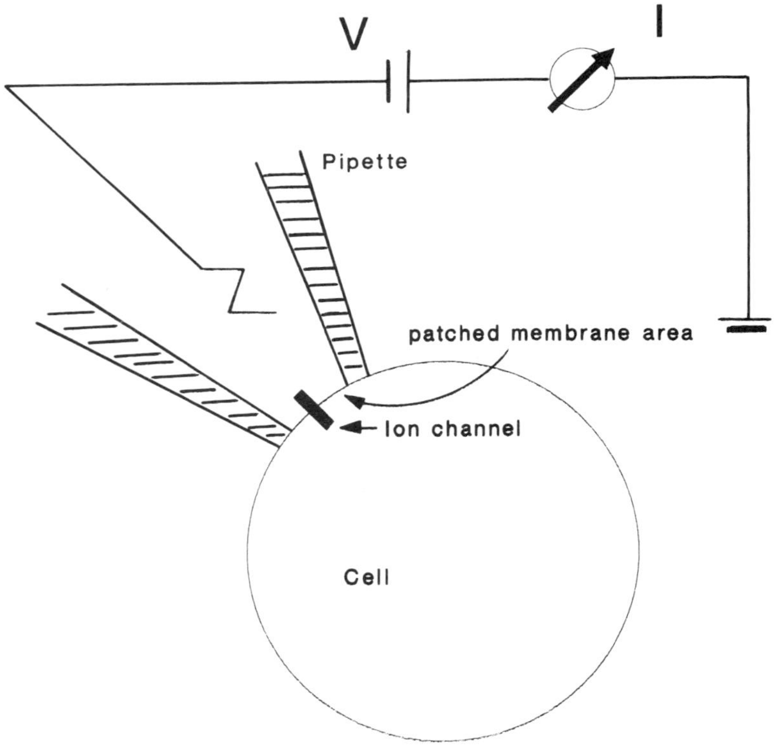


**Figure 5.7** Principles of the configuration of a patch-clamp experiment. A part of a biological membrane from a cell is electrically isolated with a glass pipette and then investigated. The membrane patch may be drawn away from the cell to establish a cell-free patch and to access an isolated small membrane area. For details, see at the Introduction and the literature.

just that (Coronado and Latorre, 1983; Schuerholz and Schindler, 1983; Suarez-Isla *et al.*, 1983; Wilmsen *et al.*, 1983; Hanke *et al.*, 1984). The resulting bilayers are frequently called 'dip-stick bilayers'.

Another obvious approach is to make liposomes so large that they can be visualized under the microscope. Then these liposomes can directly be patched, identical to any other patch-clamp experiment. This type of experiment is identical technically with a patch-clamp experiment and therefore will be discussed only shortly at the end of this chapter together with some general remarks. However, there are other even more simple ways to make planar bilayers on tips of glass pipettes, which will be introduced now.

The basic requirement of dip-stick bilayer experiments is to have appropriate glass pipettes. The pipettes used are identical to those for patch-clamp experiments. The pipettes are made from capillary glass with an electrode puller. Tip-diameters should be in the micrometre range for bilayer experiments. A further processing of the pipettes, for example fire polishing and coating, as done for some patch-clamp experiments, is not necessary. The pipettes should be used immediately after pulling. More details on making patch-pipettes are given in Chapter 3.

### 5.3.1 Simple method to make a dip-stick bilayer

The simplest way to make a dip-stick bilayer is as follows:

1. A pipette tip is immersed in an aqueous solution just below the surface (pipette mounted on a good micromanipulator, see Chapter 3 for technical details).
2. To the water surface, close to the tip of the pipette, a drop of lipid dissolved in hexane is added. The tip of the pipette is covered by this lipid-solution.
3. After waiting a few minutes, suction is added to the interior of the pipette. This can best be done by using a commercial pipette holder, which enables the application of suction, as widely used in patch-clamp experiments (see Appendix 4 for dealers). When applying suction to the drop at the tip of the pipette, the lipid is pulled up in the pipette and thinned out. Finally the state of a planar lipid bilayer is reached.

In Figure 5.8 a sketch of the technique is given. As no optical control of the bilayer is necessary and a microscope is often not available, the formation of the bilayer should be monitored by capacity measurement. In fact one of the advantages of these techniques over patching large liposomes is that no microscope, which is quite expensive, is needed. A diagram of a capacity recording of such a bilayer experiment is given in Figure 5.9. After formation of the bilayer, the experiment may proceed as for all types of bilayers. A drawing of the equipment absolutely necessary for dip stick experiments as described above has already been given in Figure 3.15. Nevertheless, a good patch-clamp set-up is

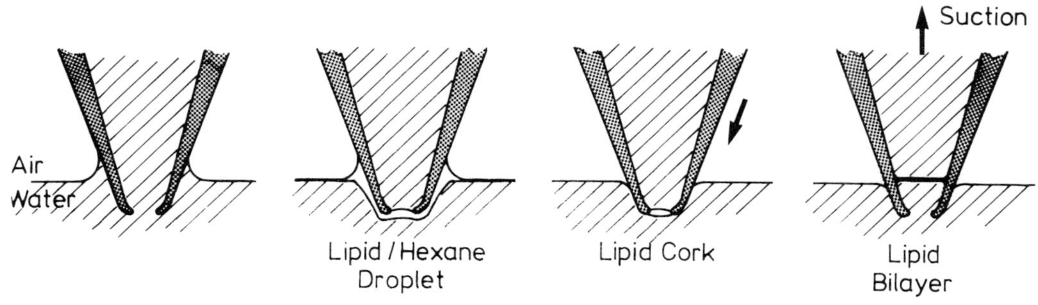


**Figure 5.8** Formation of bilayers on the tip of glass pipettes from a drop of lipid dissolved in a non-polar organic solvent (hexane). The lipid is added to the tip of the pipette when this is below the air–water interface. Using suction applied via an appropriate pipette holder (see Chapter 3) the lipid is pulled up into the pipette, thinned out and a bilayer formed. On a patch-clamp set-up this procedure can be monitored through the microscope as the lipid–water boundary can be visualized under phase contrast.

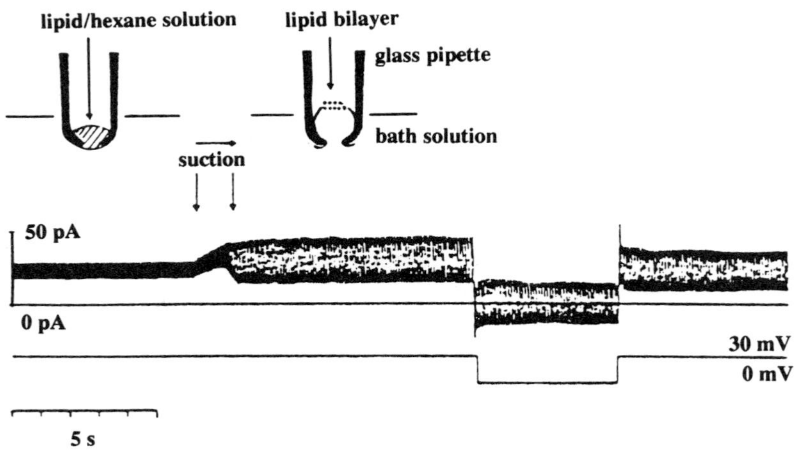


**Figure 5.9** Capacity recording during formation of a bilayer on tip of a glass pipette. When suction is applied, the capacity becomes higher, indicated by current amplitude measurement with a sine-wave potential applied to the membrane. After thinning out and after formation of the real bilayer, the seal and the conductance of the bilayer are tested by an additional voltage pulse. (Reproduced from Wilmsen *et al.*, 1983, with permission.)

fine instrumentation for dip-stick experiments and this is another advantage of the technique, as many electrophysiological laboratories now have patch-clamp equipment available. If one wants to built a simpler device, of course the microscope and some other parts can be omitted, but the principles remain the same.

Instead of adding the lipid to the air–water interface when the pipette tip is below the water surface, it can be added direct to the pipette tip outside the aqueous solution. To avoid drying of the drop, the pipette tip has to be immersed into the water very quickly after addition of the lipid. The immersion of the pipette can best be done by a micromanipulator (see Figure 3.15).

Both techniques described here are not really dip-stick techniques, but just tricks to seal the tip of a glass pipette by a drop of lipid from which a bilayer is made. The real dip-stick technique will be described in the following.

### 5.3.2 Advanced methods to make dip-stick bilayers

The above techniques to make bilayers on tips of glass pipettes work well. They give a good stability and a good signal-to-noise ratio in current measurements. Nevertheless they have the disadvantage that organic solvent is present in the system again. This may be avoided by using a more advanced procedure:

1. At an air–water interface a monolayer is built as described for the technique of folded bilayers (form a bulk solution of lipid in a highly volatile solvent or from liposomes). When the monolayer is established from a vesicle suspension it is solvent free.
2. After formation of the monolayer, a glass pipette is repetitively moved through the surface very smoothly by use of a micromanipulator. Finally there will be monolayers at the air–water interfaces at the tip of the pipette when it is outside the bath solution, and on the top of the bath solution.
3. Now the pipette has to be moved very carefully to the interface of the bath solution. By bringing the two monolayers into contact a bilayer is formed. This movement of the pipette must be done with a very good micromanipulator, for example a hydraulic one as is often used in patch-clamp experiments is suitable.
4. After contact of the two monolayers the pipette should not be moved more than a few micrometres below the surface, usually it should be held just at the surface, otherwise the stabilty of the bilayer is reduced.
5. For experiments which need perfusion of the aqueous solution bathing the bilayer, the pipette can instead be moved below the surface to a perfusion pipette. Details of such a perfusion system are given in Figure 5.10 (the system is identical to perfusion systems in patch-clamp experiments, consequently the

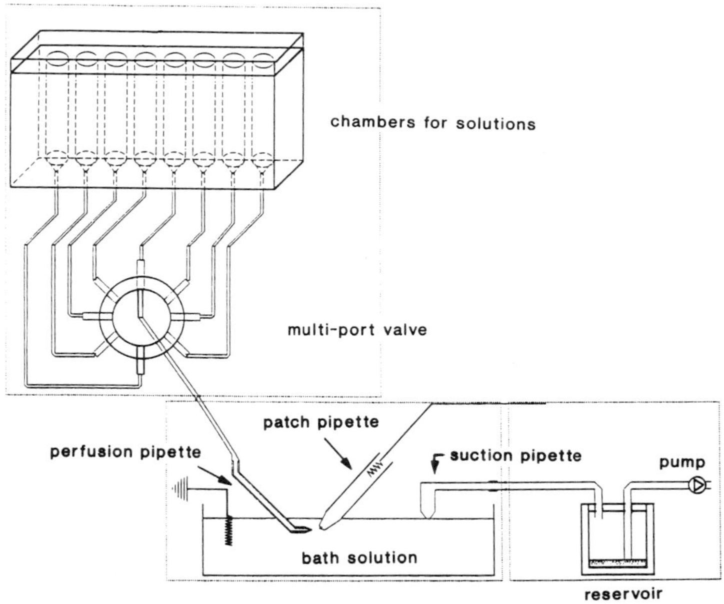


**Figure 5.10** Perfusion system for dip-stick bilayers. The same system can be used for patch-clamp experiments. A perfusion capillary, diameter about 100 μm, is connected to a multi-port valve. Each port of the valve is connected to a bottle of solution. The solution flows by gravitation when the bottles are placed higher than the tip of the capillary. The bilayer pipette is brought close to the perfusion pipette via the micromanipulator (a microscope is necessary). By switching the valve, the bilayer can be perfused with different solutions. Exchange time usually is in the range of seconds up to a minute, depending on details of construction. In addition, a pump is needed to suck away the aqueous solution from the bath chamber.

types of equipment used in the latter can be used with dip-stick bilayers).

It is obvious that mechanical movement of the surface of the water is deleterious to making bilayers with the dip-stick technique. Therefore really good vibration isolation of the work bench is necessary. Additionally the evaporation of the water, lowering the surface level on a longer time scale, has to be taken into account. If necessary, the pipette may be readjusted after some time to prevent loss of contact.

In Figure 5.11 a diagram of the technique is shown. Again the formation of the seal and the bilayer is best monitored by capacity measurement as shown in Figure 5.11. The bilayers built with this techniques are really solvent free. In addition, their areas are quite small, giving a good signal-to-noise ratio in electrical current recording. It has been found experimentally that this type of bilayer is very stable even at high membrane potentials. So there are a lot of advantages over classical bilayers, although, as usual, disadvantages have to be taken into account. First, access to the inside of the pipette is very difficult (for example in perfusion experiments); second, the small area reduces the rate of incorporation of protein molecules in reconstitution experiments and last, but not least, there is a high rate of breaking of bilayers when trying to form them or to handle them in perfusion experiments as shown in Figure 5.10. A final point we would like to address here is the finding that making bilayers from lipid mixtures with this technique may result in the formation of asymmetric membranes on the tip of the pipette, a fact which has been described in the literature (Coronado and Latorre, 1983) as a consequence of enhanced lipid flip-flop during bilayer formation; this finding is poorly understood to date and can not be controlled in the experiments. The same should hold for patch-clamp experiments but is usually neglected as the spontaneous lipid flip-flop cannot be measured or controlled with today's techniques.

The structure of the membrane at the tip of the pipette is still unclear; there are some ideas, however, about what may happen. The surface of the glass of the pipettes is hydrophilic, by this the lipids must expose their hydrophilic head-groups to the glass. The consequences of this are illustrated in Figure 5.12. There, a diagram of the structure of a pure lipid bilayer without any solvent at the tip of a glass pipette is given.

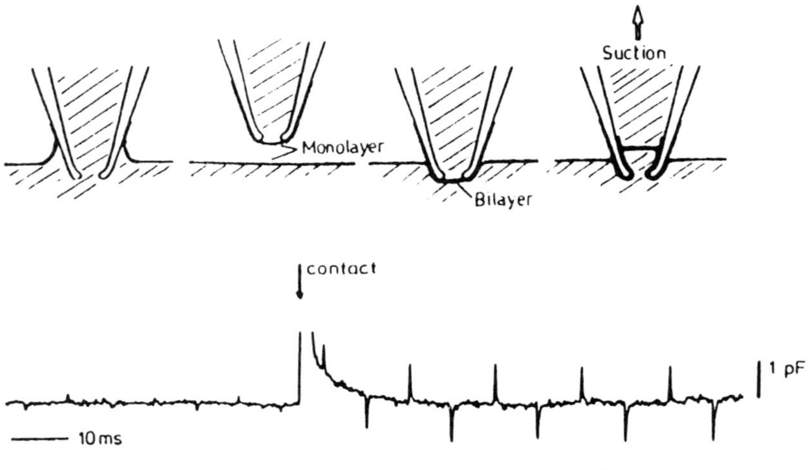


**Figure 5.11** Formation of a bilayer on the tip of a glass pipette by contact of two monolayers. After a monolayer has been formed, the pipette is moved out of the water and then the two monolayers which are established at the air–water interfaces are brought to contact. Some suction may be applied to the interior of the pipette to stabilize the bilayer. The capacity recording shows the formation of the seal and the bilayer at the moment of contact. In this technique, at least for moving the pipette up and down, a high quality microdrive is essential. (Reproduced from Hanke *et al.*, 1984, with permission.)

Finally, we want to point out a more sophisticated technique to make dip-stick bilayers, which was introduced by Schindler and co-workers (Schuerholz and Schindler, 1983). This technique brings us back a step to the idea of Montal and Mueller (and Takagi *et al.*, 1965) on making bilayers from monolayers of controlled surface pressure. Again the bilayer is made from a monolayer on an air–water interface. First the monolayer at the surface of the bathing solution is made on a small film balance, giving the required physical control of this part of the membrane. The monolayer at the tip of the glass pipette is connected to another film balance by guided contact over wet paper, thus providing a partial control of its lateral preasure. After formation of the two monolayers they are brought together by a micromanipulator as is done in the simple dip-stick technique. A diagram of the set-up is given in Figure 5.13; here in addition the structure of the type of bilayer formed is shown. It is obvious that there is no direct connection beween the monolayers and the bilayer because of the boundary of the glass, which will prevent a perfectly controlled change of the surface pressure of the monolayers at the tip of the glass pipette. Nevertheless, with such a set-up, for the first time the lateral pressure of the monolayers a bilayer is made from could be changed during the experiment, and the effect of this on the bilayer properties can be studied.

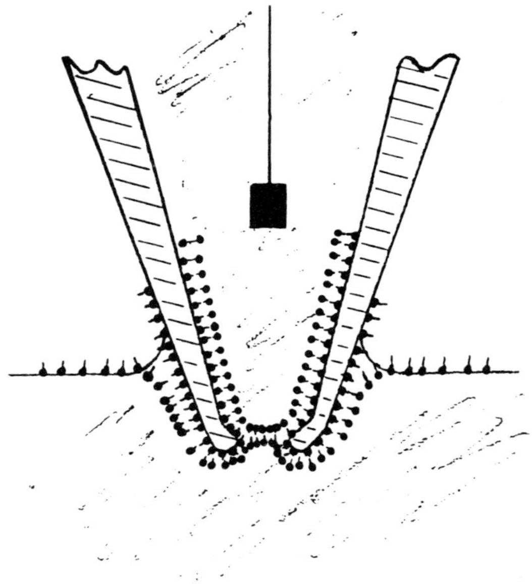

**Figure 5.12** Simplified structure of a bilayer at the tip of a glass pipette. The coating of the hydrophilic glass by lipid demonstrates the needs of stable bilayers.

As we will present a set-up for making bilayers under really controlled conditions in Chapter 8, we will not focus much on technical details about this type of dip-stick bilayer here. It has the disadvantage (identical to the example given in

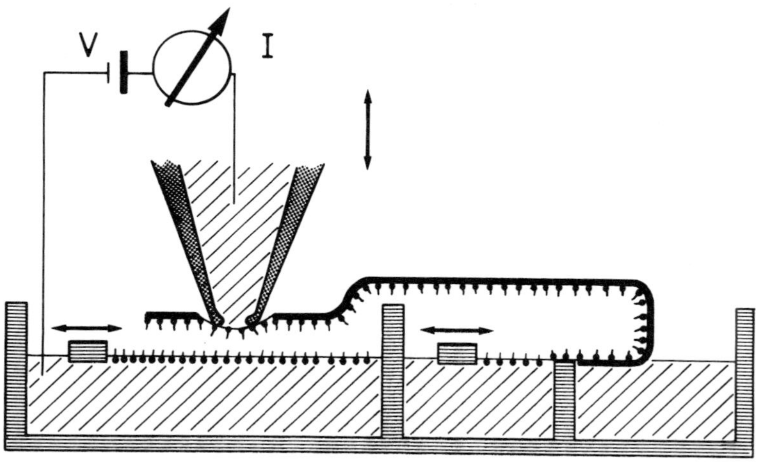


**Figure 5.13** The formation of bilayers on the tip of a glass pipette by guided contact of two monolayers. Here the monolayers are formed on film balances. The lateral pressure of the monolayers can be controlled as in classical measurements with film balances. At the contact of the lipid with the wall of the pipette, the electrical seal is formed, but additionally the continuity of the monolayer is broken. There is no direct contact between the two monolayers and the bilayer. Therefore a change of lateral pressure in the monolayers will not necessarily directly affect the bilayer (see also Schuerholz and Schindler, 1983).

Chapter 8) that a complete new set-up has to be built and that no patch-clamp set-up can be used. As stated already, this is just one of the main advantages of the other dip-stick techniques. Nevertheless here are a few details about how to perform an experiment with a dip-stick bilayer from monolayers on film balances for people who are interested in trying it.

First, the areas of the film balances for the monolayers should be small; a surface area of 10 $cm^2$ for each chamber is sufficient. The film balance for the monolayer on the tip of the pipette may be even smaller. The two film balances are best placed in one mechanical block made from Teflon. From the film balance used for the pipette tip, a small piece of wet paper should lead to the tip of the pipette itself. It must be carefully fixed to the film balance chamber to give good contact for the continuous monolayer. To best contact the tip of the pipette, a small hole is made in the paper in which the tip of the pipette is immersed. The paper and the pipette must be fixed by appropriate mechanical equipment. In the hole there will be a small drop of water, which continues to the tip of the pipette when this is adjusted properly. After formation of both monolayers, these are brought together by moving the pipette downwards with a good micro-manipulator. A one-axis microdrive can be used as the pipette has to be moved carefully only in one direction. The paper should of course be moved together with the pipette. After contact of the two monolayers, a bilayer is formed and the experiment can proceed as for other dip-stick experiments. The lateral pressure of the monolayers should be between 30 dyne/cm and 45 dyne/cm as was described by Schindler (Schindler and Feher, 1976; Schindler, 1980; Schuerholz and Schindler, 1983). To measure the lateral pressure of the two monolayers (if required), a 'Wilhelmy' type of probe as described later in the book (Chapter 8), would be appropriate. The main problem with this technique is to properly adjust the paper at the tip of the pipette and at the film balance to give a perfect monolayer from the surface of the film balance to the tip of the pipette. Due to the difficulties encountered in handling, this technique has not become as common in bilayer and patch-clamp laboratories as other dip-stick techniques.

## 5.4 PATCH-CLAMPING LARGE LIPOSOMES

Quite often at the end of a biochemical procedure a preparation of vesicles containing proteins in their membranes is produced. If these vesicles can be made large enough – the average diameter should be greater than 5 µm to visualize them under a good microscope – they can be directly used in electrophysiological experiments by application of the patch-clamp technique. This procedure is identical to a classical patch-clamp experiment, except instead of a cell a large vesicle is used (Tank and Miller, 1983; Keller *et al.*, 1988). Besides the fact that large liposomes are necessary, there are no other requirements than those of a patch-clamp experiment. Different techniques to make these large liposomes have been described in Section 4.3 in detail. We will not describe the patch-clamp technique here, but just refer to the literature available and show in Figure 5.14 a diagram of a patched large liposome in the cell-attached configuration. Additionally in the lower part of this Figure a cell-free patch is shown. As indicated at the top of the Figure, such an experiment starts with a preparation of purified protein (channel-forming protein) to be inserted into the membrane of artificial lipid vesicles.

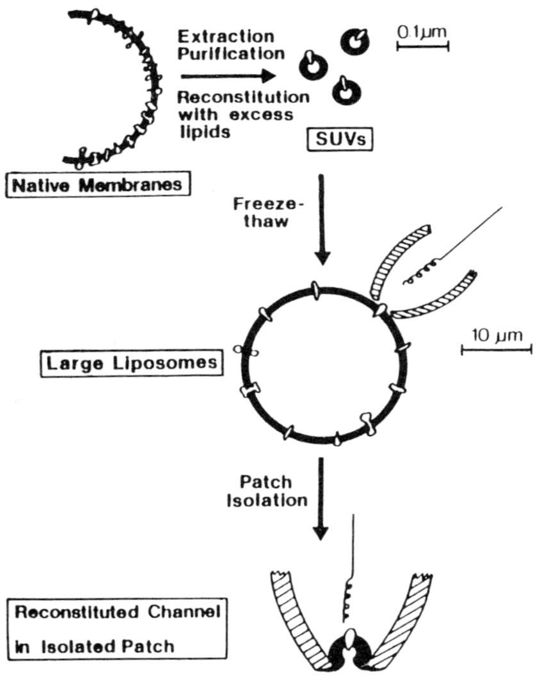


**Figure 5.14** A large vesicle patched by a glass pipette. The experiment starts with a preparation of purified protein in the membranes of artificial lipid vesicles (top). First the cell-attached configuration is established, then, in the lower part, the patch pipette is shown in the cell-free configuration. (Reproduced from Tank and Miller, 1983, with permission)

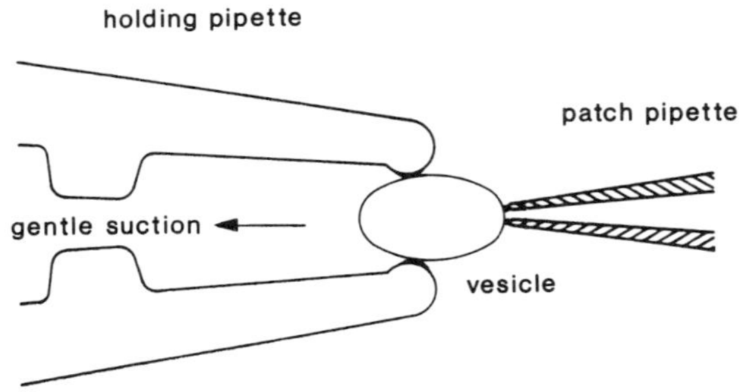


**Figure 5.15** A vesicle immobilized by a large holding pipette and patched by a glass micropipette. For this procedure two adequate micromanipulators are necessary. The described procedure may be used to avoid drifting away of vesicles, which are not adhering to the bottom of the culture dish. This is of special interest for excising patches to form cell-free patches or for use with a perfusion system.

The technique of patching large liposomes combines two complementary techniques, adding a few advantages of bilayer experiments to all the advantages of the classical patch-clamp technique. The advantages of the bilayer are:

1. choice of lipid composition
2. use of purified protein preparations
3. access to systems which cannot be patched directly
4. control of physical state of the membrane.

A few experimental aspects which are different from the standard patch-clamp technique have to be taken into account to perform successful experiments with patched large vesicles. Usually the vesicles are not immobilized by sticking to the bottom of the culture dish but are just added to the bath solution after formation (sometimes vesicles spontaneously stick to the bottom of plastic dishes, however). The freely-moving vesicles are no real problem during a patch-clamp experiment. One has only to be careful in approaching the vesicle. A positive pressure which sometimes is applied to the patch-pipette to prevent its tip from contamination must not be too big. For perfusion of the system, however, immobilization of the vesicle may be useful and can be done by a second, bigger, holding-pipette as shown schematically in Figure 5.15.

Some of the procedures to make large vesicles as described in Section 4.3 result in large unilamellar vesicles. The membranes of these liposomes can be handled identically to plasma membranes. Other procedures, however, preferentially give large multi-lamellar liposomes (the freeze–thaw technique for example). Although they can be handled identically, sometimes the yield of successful patches is better, when patching so-called blebs of these liposomes. These blebs occur spontaneously at a lower rate and can be produced by slight hypo-osmotic treatment of the vesicles. They are regions of mono-membrane structure and therefore resemble monolamellar vesicle surfaces. To obtain cell-free patches, it is best to pull the pipette away from the vesicle (in cases of vesicles not immobilized this is best done in one go) and briefly expose its tip to air. This procedure reproducibly results in inside-out patches from the vesicle membrane which then can be used as in patch-clamp experiments, independent of any structure complexity of the vesicle at the beginning.

At the end of this chapter it seems to be useful to ask the question, which bilayer technique is the 'best' one? Of course this question can not be answered simply, as any answer would depend on the particular experimental problem to be solved. However, we want to make a few statements which may help the inexperienced to select the technique for experiments:

1. For learning the basics of bilayer experiments, the technique of painted bilayers seems to be suitable as it is technically quite easy to handle.
2. For investigation of artificial pore-forming polypeptides and many channel-forming proteins the same holds, especially as in reconstitution experiments the technique of painted bilayers has some advantages due to the possibility of more easily inserting proteins (see Chapter 6).
3. For reconstitution of critical ion-channel forming proteins the technique of folded bilayers is adequate.
4. For experiments investigating the physical properties of membranes, folded bilayers may be preferred, as they contain much less solvent.
5. For experiments that require high current and time resolution, all pipette (dip-stick) techniques are suitable.
6. For fast screening of biological probes in bilayer experiments the pipette (dip-stick) techniques have some advantages, too.
7. In laboratories with a patch-clamp set-up available the use of pipette (dip-stick) techniques has the advantage that no additional set-up is needed.

Of course, the above statements are the preferences of the authors. Other opinions are possible. However, this list may at least be an aid to finding the relevant approach to an experimental problem in creating bilayer experiments.

## REFERENCES

Adamson, A.W. (1976) *Physical chemistry of surfaces*. 3rd edit., Wiley, New York.

Coronado, R. and Latorre, R. (1983) Formation of phospholipid bilayers on patch-clamp pipettes. *Biophys. J.* **43**, 231–236.

Duschl, C., Kemper, D., Frey, W., Meller, P., Ringsdorf, H. and Knoll, W. (1989) Relation between phase diagram, crystallization and optical properties of dye/stearic acid mixed monolayers. *J. Phys. Chem.* **93**, 4587–4593.

Fettiplace, R., Gordon, L.G.M., Hladky, S.B., Requena, J., Zingsheim, H.P. and Haydon, D.A. (1975) Techniques in formation and examination of black lipid bilayer membranes. In: *Methods of membrane biology*, Vol. 4 E.D. Korn (ed.), Plenum Press, New York, pp. 1–75.

Finkelstein, A. (1974) Bilayers: Formation, measurements and incorporation of components. In: Methods of enzymology, Volume XXXII, Biomembranes, Part B., S. Fleischer and L. Packer, (eds.), Academic Press, New York, pp. 489–501.

Hamill, O.P., Marty, A., Neher, E., Sakmann, B. and Sigworth, F.J. (1981) Improved patch-clamp technique for high resolution current recording from cells and cell-free membrane patches. *Pflügers Arch.* **391** 85–100.

Hanke, W., Methfessel, C., Wilmsen, U. and Boheim, G. (1984) Ion channel reconstitution into planar lipid bilayers on glass pipettes. *Biochem. Bioeng. J.*, **12**, 329–339.

Hanke. W. (1985) Reconstitution of ion channels. *CRC Crit. Rev. Biochem.* **19**, 1–44.

Jain, M. (1972) *The bimolecular lipid membrane, a system*. Van Nostrand Reinhold, New York.

Kaufmann, K. (1989) *Role of lipid membrane*. Caruaru, Brazil.

Keller, B.U., Hedrich, R., Vaz, W.L. and Criado, M. (1988) Single channel recordings of reconstituted ion channels: an improved technique. *Pflügers Arch.* **411**, 94–100.

Miller, C. (1986) *Ion channel reconstitution*. Plenum Press, New York.

Montal, M. and Mueller, P. (1972) Formation of bimolecular membranes from lipid monolayers and a study of their electrical properties. *Proc. Natl. Acad. Sci. USA* **69**, 3561–3566.

Mueller, P., Rudin, D., Tien, H.T. and Wescott, W.C. (1962) Reconstitution of excitable cell membrane structure in vitro. *Circulation* **26**, 1167–1171.

Sakmann, B. and Neher, E. (1983) *Single channel recording*. Plenum Press, New York.

Schindler, H. (1980) Formation of planar lipid bilayers from artifical or native membrane vesicles. *FEBS Lett.* **122**, 77–79.

Schindler, H. and Feher, G. (1976) Branched molecular lipid membranes. *Biophys. J.* **16**, 109–1113.

Schuerholz, T. and Schindler, H. (1983) Formation of lipid-protein bilayers by micropipette-guided contact of two monolayers. *FEBS Lett.* **152**, 187–190.

Suarez-Izla, B.A., Wan, K. Lindstrom, J. and Montal, M. (1983) Single channel recordings from purified acetylcholine receptors reconstituted in bilayers formed at the tip of patch pipettes. *Biochemistry* **22**, 2319–2323.

Takagi, M., Azuma, K. and Kishimoto, U. (1965) A new method for the formation of bilayer membranes in aqueous solution. *Ann. Rept. Biol., Osaka University* **13**, 108–114.

Tank, D.W. and Miller, C. (1983) Patch-clamped liposomes: Recording reconstituted ion channels. In *Single-channel recording*. B. Sakmann and E. Neher, (eds.), Plenum Press, New York, pp. 91–106.

Ti Tien, H. (1974) *Bilayer lipid membranes*. Marcel Dekker, Inc., New York.

van den Berg, H.J. (1965) A new terchnique for obtaining thin lipid films seperating two aqueous media. *J. Mol. Biol.* **12**, 290–296.

White, S.H. (1986) The physical nature of planar bilayer membranes. In: *Ion channel reconstitution*. C. Miller, (ed.), Plenum Press, New York, pp. 3–35.

Wilmsen, U., Methfessel, C., Hanke, W. and Boheim, G. (1983) Channel current fluctation studies with solvent free planar lipid bilayers using Neher–Sakmann pipettes. In *Physical chemistry of transmembrane ion motion*. G. Spach (ed.), Elsevier, Amsterdam, pp. 479–485.

CHAPTER SIX

# Incorporation of proteins into planar lipid bilayers

At the beginning of bilayer work in the 1960s two problems were investigated almost exclusively, the electrical properties of pure lipid membranes and the ion-transport properties of some simple, 'artificial' transport systems incorporated into planar lipid bilayers. These transport systems were usually isolated, purified (later synthetic) polypeptides of limited length (some 10–20 amino acids). Some other compounds which were water- as well as lipid-soluble and therefore quite easy to incorporate into planar lipid bilayers (see later), have also been investigated. Simple carriers as well as ionic pores were studied by these experiments in some detail (Ti Tien, 1974). A few well-known examples which are still of interest today due to their model character, are gramicidin (Hladkey and Haydon, 1984; Woolley and Wallace, 1992), alamethicin (Boheim, 1974) and valinomycin (Benz and Läuger, 1976). Alamethicin will be discussed in more detail, as it is a good simple model-system for studying structure-function relations of pore forming proteins (Boheim, 1974). Even with today's molecular biological methods, such studies can be done only to a very limited extent with the highly complicated molecules that form pores in biological membranes (Anderson and Koeppe, 1992). Furthermore alamethicin is well suited to test the bilayer set-up. Additionally, the investigation of substances like alamethicin or gramicidin in bilayer research is still of interest in the investigation of drugs and toxins in pharmacological studies (Tamm, 1991).

However, the investigation of isolated membrane transport systems from biological membranes after incorporation into simplified artificial systems (e.g. membranes of defined biochemical composition and easy access to both sides of the membrane) is one of the advantages of planar lipid bilayer experiments. Therefore, starting in the 1970s, methods were developed to isolate membrane proteins (see chapter 4) and to incorporate them into planar lipid bilayers (Darzon, 1986; Hanke, 1985; Latorre, 1987; Miller, 1983, 1986).

In the previous chapter we have discussed how

to make planar lipid bilayers as a prerequisite of protein incorporation. In Chapter 4 we presented techniques of protein and lipid biochemistry to obtain the necessary preparations. In this chapter we want to proceed to the next step, to introduce techniques to incorporate polypeptides and proteins into planar lipid bilayers. Here we will proceed from simple techniques of incorporating water-soluble polypeptides into planar bilayers, to more complicated questions such as the incorporation of integral membrane proteins into model membranes. Additionally, we will discuss an approach to attaching membrane fragments to pre-formed bilayers, as this is a quite useful approach for some special questions in the investigation of ion carriers (although it is not a real incorporation of proteins into a lipid bilayer).

For technical reasons the handling of dip-stick bilayers and incorporation techniques will be dicussed separately. Furthermore, in the following we will not discriminate between proteins and polypeptides, but will focus on their common properties relevant to incorporation techniques.

## 6.1 WATER-SOLUBLE COMPOUNDS

A large number of membrane transport systems, including ion carriers, pore-forming polypeptide-antibiotics and toxins are water- as well as lipid-soluble. Usually these substances can be held in aqueous stock solutions of moderate concentration. Sometimes a small amount of alcohol (or other organic solvent) is used to increase the solubility of certain drugs.

The incorporation of these substances into planar lipid bilayers is quite straightforward. A bilayer can be made according to any of the techniques which have been described in Chapter 5. To the aqueous solutions at either sides of the bilayer (or both) a small amount of stock solution of the material to be incorporated into the bilayer can be added. After that, the solution to which the material has been added is stirred vigorously to shorten mixing time (stirring of course has to be stopped when, for example, recording of current across the bilayer begins). Part of the material will be incorporated into the bilayer by diffusion, regulated by the solubility of the material in lipid and in the aqueous phase. After waiting an adequate period of time the measurements of the properties of the incorporated molecules can be initiated.

There are a few things one should bear in mind when adding drugs to the bath solution. First, incorporation into the bilayer is concentration dependent, but additionally, and more importantly the incorporation is dependent on the distribution coefficient of the added compound between water and lipid (partition coefficient). This will affect the final concentration of the compound in the bilayer. Thus the final conductivity of the bilayer and the time which is needed to reach equilibrium is dependent on time, drug concentration and the partition coefficient.

Experimentally, it is best to measure the incorporation of any drug into the bilayer as a function of time and bath concentration before starting other investigations. From this, the time interval necessary to approach an equilibrium concentration in the bilayer can be determined. This interval can be quite long for some drugs, as has been shown for example for gramicidin. Up to ten minutes (sometimes even longer) were found to be necessary to reach a conductance level that was stable in time, after adding gramicidin to the aqueous solution bathing a bilayer.

As mentioned above, sometimes low concentrations of organic solvent (e.g. alcohol) are added to the stock solution of drugs to increase solubility. When using such solutions, the effect of the solvent in its final concentration on the bilayer should be studied in control experiments. It is known, for example, that alcohol and chloroform increase bilayer conductance.

### 6.1.1 Detergent solubilized compounds

Proteins and other organic compounds which are not water soluble can be made water soluble by adding detergent to the aqueous solution. In fact this method is used as a starting point for extraction of membrane-bound proteins from the membrane itself, and for this reason many biochemical procedures to isolate or purify

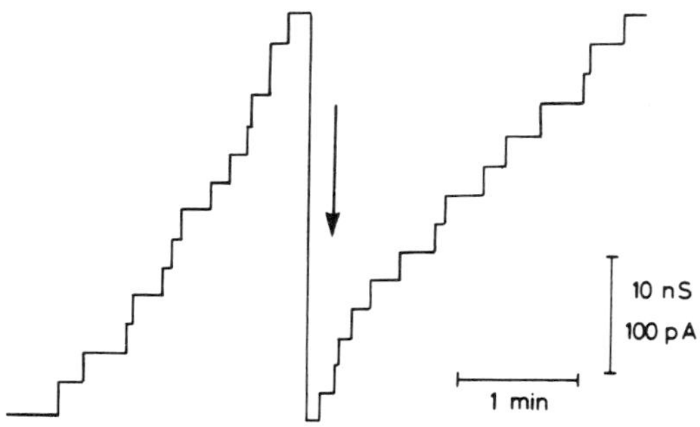


**Figure 6.1** Incorporation of matrix protein from *E. coli* into a planar lipid bilayer. The protein was solubilized by use of detergent. A painted bilayer was made at the beginning of the experiment and the protein was added to one side of the bilayer (aqueous solution contained 1 M KCl). The probability of the channels formed by the protein being in the open state is close to one, thus the stepwise increase of conductance indicates the incorporation of single ion channels into the bilayer. (Reproduced from Benz *et al.*, 1986, with permission)

membrane transport proteins include steps of dissolving the protein in an aqueous solution with moderate concentrations of detergent.

This aqueous solution of protein made soluble by detergent can provide the starting material for incorporation of a protein into a bilayer. The material can be handled the same way as described above for water-soluble compounds, i.e. the solution can be added directly to the bath solution of the bilayer. In Figure 6.1, as an example, the incorporation of detergent-solubilized matrix protein from *Escherichia coli* into planar lipid bilayers is shown (Benz *et al.*, 1986). As the open-state probability of the channels formed by this protein is close to one, the channels being almost permanently open, thus the incorporation of each functional channel into the bilayer can be seen as a step-like increase in the current recording.

Again a problem which has to be taken into account is the effect of the detergent itself on the bilayer. Most of the detergents at low concentration induce some baseline noise and a non-specific increase in conductance in planar lipid bilayers. However, it has been found that ion-channel like fluctuations can be observed after addition of some detergent types to pure lipid bilayers. In any case, the effect of the concentration of detergent used in the experiments on the bilayers should always be studied carefully in control experiments.

At high concentrations of detergent the stability of the bilayer is usually decreased. This can be seen in the spontaneous occurrence of non-specific fluctuations and breakdown of the membrane. In such cases either the detergent concentration should be reduced or other incorporation techniques should be used.

Finally, we want to point out that adding a small amount of stock solution with detergent to a large volume of bath solution without detergent, may itself raise a problem. The concentration of the detergent may fall below the value necessary to keep proteins in solution. This might result in denaturation of the protein.

As a consequence, the final concentration of detergent in the bath solution should be carefully calculated, so that a compromise between too high and too low concentration must be found.

### 6.1.2 Compounds dissolved in organic solvent

A method which has sometimes been used in bilayer experiments is to directly add a compound that is of interest but is not water-soluble (protein, etc.), to the organic solution of the lipid from which the bilayer is made. Either the material is directly added to the lipid stock solution, or a solution of material in another organic solvent is added to the lipid stock solution. Of course this can only be done in bilayer techniques using lipid dissolved in organic solvents as the starting material, for example, the methods of painted bilayers and of folded bilayers. Methods using lipid in the form of vesicles are not possible here. Following the ideas outlined earlier, the effects of the additional organic solvents used on the bilayer must be studied in control experiments.

A general question of interest, along with this section, is the effect of the organic compounds and detergents on the properties of the transport systems under investigation. This is in principle identical to the question of whether the solvent in which the lipid is dissolved affects the transport system reconstituted. As an example of how significant these effects can be, it should be mentioned here that the nAChR has never been successfully reconstituted into painted bilayers. Because comparable experiments with folded bilayers (in which all parameters other than the bilayer type were identical) were successful, this can only be a result of the larger amount of organic solvent in the painted bilayer than the folded bilayer experiment. As a consequence, control experiments with other techniques are necessary when the incorporation techniques discussed above are used. An ideal situation, for example, is given when data for the transport system under investigation are available from patch-clamp experiments (*in situ*) or from *in vivo* experiments.

## 6.2 MEMBRANE PROTEINS

In contrast to the techniques discussed in Section 6.1, most of the transport systems from biological membranes are integral membrane proteins and therefore not water-soluble. Usually they cannot be dissolved in organic solvents due to the resulting denaturation of the protein. The technique of making proteins water soluble by use of detergent and then adding them directly to the aqueous phase of the bilayer set-up is in most cases not the technique of first choice, due to unknown effects of the detergent on the function of the transport system and the bilayer itself. Finally no biological membranes contain detergent, and thus membranes containing detergent are not comparable with biological membranes. As a result other techniques have to be used, starting with the material which results from most biochemical preparations: liposomes with the transport proteins in their membranes. Two techniques have been available to date and are discussed here.

### 6.2.1 Spreading of protein-containing liposomes at air–water interfaces

As described earlier, Schindler and co-workers (Schindler, 1982, 1989) introduced a modified technique of making folded bilayers. The monolayers they made their bilayers from were spread from liposomes. These monolayers are in equilibrium with the liposomes in the bath solution and can be quite easily established from different liposome preparations. One approach to this technique is to use liposomes that have the proteins to be reconstituted already in their membranes. When spreading such liposomes at an air–water interface not only does the lipid spread but so do the proteins in the liposome membranes. After having reached equilibrium at the air–water interface, a stable film being composed of lipid and protein (somehow reflecting the stoichiometry of the vesicles) should be established at the surface. By combining the two monolayers on each side of a partition into a bilayer as described in Chapter 5, a bilayer is formed which should have the protein already incorporated. The composition of the bilayer to some extent will reflect the composition of the liposomes from which the monolayers were spread. All the technical details here are the same as for simple bilayer production by spreading liposomes at an air–water interface as already described. Figure 6.2 shows all the steps necessary in the reconstitution of proteins into planar lipid bilayers by spreading protein containing

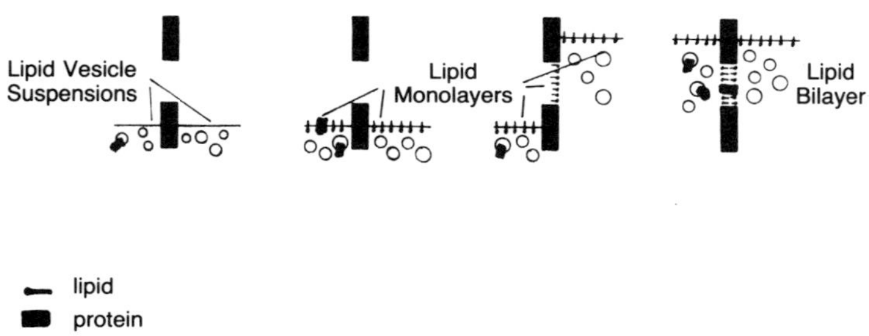


**Figure 6.2** Incorporation of proteins into planar lipid bilayers by spreading protein-containing liposomes at an air–water interface. Apart from the fact that protein containing liposomes are used, the technique is identical with the technique of making folded bilayers from spread liposomes as shown in Chapter 5 in more detail.

liposomes at an air–water interface (see Chapter 5 for comparison).

The technique of spreading protein-containing liposomes has a few obvious advantages. First, the concentration of protein in the bilayer can be controlled by the amount of protein in the liposome membranes. A concentration calibration is difficult but, as stated already, there will be a functional dependence of protein concentration in the bilayer on protein concentration in the liposome membranes and on the liposome concentration in the aqueous phases. Second, bilayers can be made symmetrically as well as asymmetrically (see above), the protein can be spread with both monolayers or with only one, and thus the effect of asymmetric membrane composition on protein function can be studied. This is especially helpful when agonist-activated channels are under investigation, as it may help to have channels incorporated in the bilayer only in one orientation.

On the other hand, there are some features of this method which reduce its efficiency. As membrane proteins usually span the entire membrane it is unclear how these proteins are arranged in the monolayer at the air–water interface. For example, part of the protein molecule may be exposed to air, which is a highly artificial situation and may cause denaturation.

As shown by different groups, the technique in fact works well with a variety of ion-channel forming proteins, regardless of the remarks just made. A good example is nAChR channels from neuronal tissue of insects (Hanke and Breer, 1986), which has been successfully reconstituted by this method (and others) by the authors. Figure 6.3 displays typical results of the reconstitution of nAChR by the method described above. The data obtained were similar to data from comparable nAChRs obtained by patch-clamp experiments and from experiments involving reconstitution of the same protein in vesicle fusion-experiments (Section 6.2.2). In the bilayer experiments being discussed here, however, it has been found that the protein was incorporated in the artificial membrane only in one orientation when only spread at one monolayer. From the preparation used, it was known that the vesicles contained proteins in both orientations in their membranes. Thus the findings can only be explained when either by the spreading of the liposomes at the air–water interface a strongly asymmetric orientation of the protein is given or when part of the protein denatures at the air–water interface due to exposure of a critical part of the protein structure to the air. The same has been shown in more detail in reconstitution experiments with the nAChR from the electric organs of some fishes (Montal *et al.*, 1986).

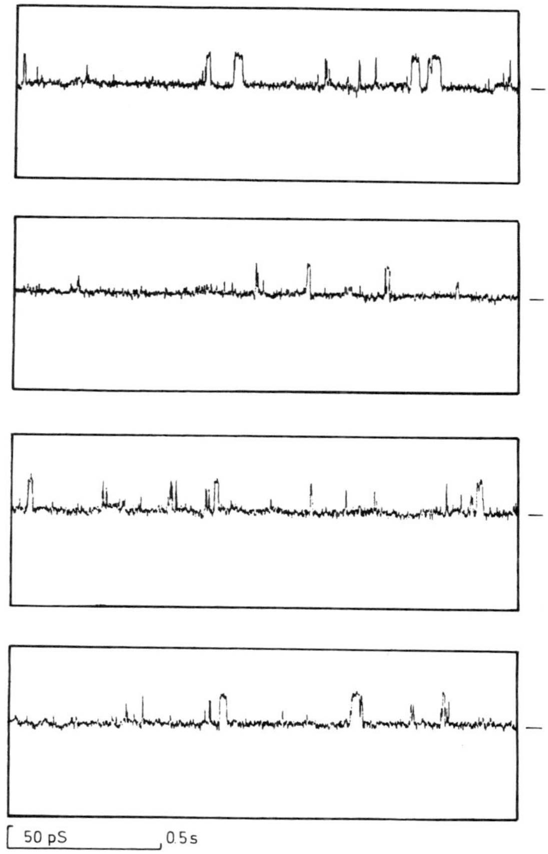


**Figure 6.3** Channel current fluctuations of reconstituted nicotinic acetylcholine receptors (nAChRs from insect neuronal tissue) in planar lipid bilayers. The protein was incorporated in the membranes of liposomes which were spread at the air–water interface of aqueous solutions to form monolayers. These monolayers were combined into a folded bilayer. Carbamylcholine was added at the vesicle side of the bilayer to activate the channels. The channel activity was measured in 200 mM NaCl at 50 mV. Single-channel parameters are similar to those known from patch-clamp experiments.

Leaving aside all the discussion about advantages

and limitations of the method, here is a procedure for an initial experiment:

1. Prepare bilayer chambers and septum as described for folded bilayers.
2. Prepare set-up and fill installed chambers below the aperture with aqueous solution, 200 mM KCl, 1mM $CaCl_2$ 5 mM Tris–HCl, pH = 7.4.
3. Use protein containing liposomes, i.e. nAChR preparation as decribed in Chapter 4, add 50 µl to the chamber on one side of the bilayer.
4. Add pure Asolectin liposomes (as described in Chapter 4), 50 µl from 20 mg/ml to the other chamber.
5. Wait 15 min.
6. Raise the level in the nAChR side of the bilayer chamber gently.
7. Raise the level in the pure vesicle side identically.
8. Add 500 µM carbamylcholine (final concentration) symmetrically to both sides of the bilayer.

The bilayer being built should have a capacity of about 0.8 $\mu F/cm^2$ and one should monitor ion-channel current fluctuations after the addition of agonist (point 8) to the bilayer, when a constant potential is applied. A more detailed description of reconstitution of the nAChR is given later.

### 6.2.2 Fusion of protein-containing liposomes with pre-formed bilayers

#### *6.2.2.1 Preliminary considerations*

As stated already (especially in Chapter 4) many membrane proteins can be incorporated into the membranes of lipid vesicles quite easily. In fact the easiest way to get a preparation of a certain membrane protein is to make a crude membrane-fragment vesicle preparation of the membranes from which the protein originates. Of course advanced techniques will proceed to the formation of vesicles from a defined lipid having a purified protein incorporated. Either of these vesicle preparations may be the starting point of another protein reconstitution technique. The vesicles containing the protein in their membranes can be fused with a pre-formed bilayer. The principles of this technique are given in Figure 6.4. A vesicle containing a protein molecule in its membrane is fused with a pre-formed bilayer, which can be made by any of the methods described earlier. By the fusion of the vesicle with the bilayer, the entire membrane contents of the vesicle become an integral part of the bilayer (Cohen *et al.*, 1980). Thus the proteins, having been incorporated in the vesicle membrane, after the fusion event, are incorporated in the planar lipid bilayer. For energetic reasons the membrane will spontaneously become flat after fusion of the liposome.

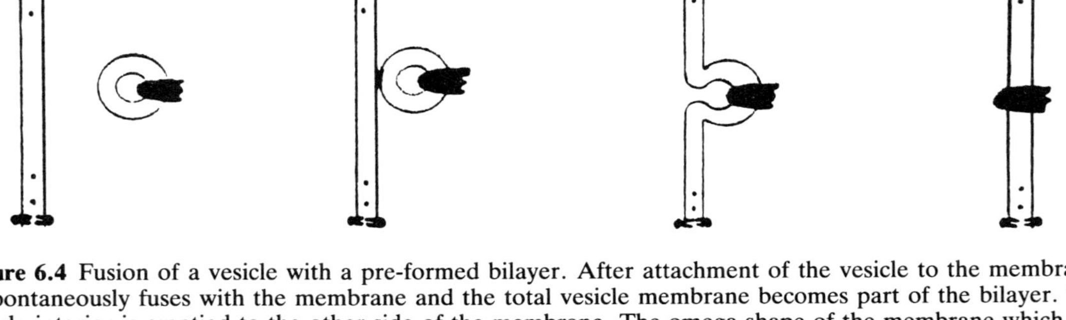


**Figure 6.4** Fusion of a vesicle with a pre-formed bilayer. After attachment of the vesicle to the membrane, it spontaneously fuses with the membrane and the total vesicle membrane becomes part of the bilayer. The vesicle interior is emptied to the other side of the membrane. The omega shape of the membrane which will exist directly after the fusion event, will become flat spontaneously for energetic reasons.

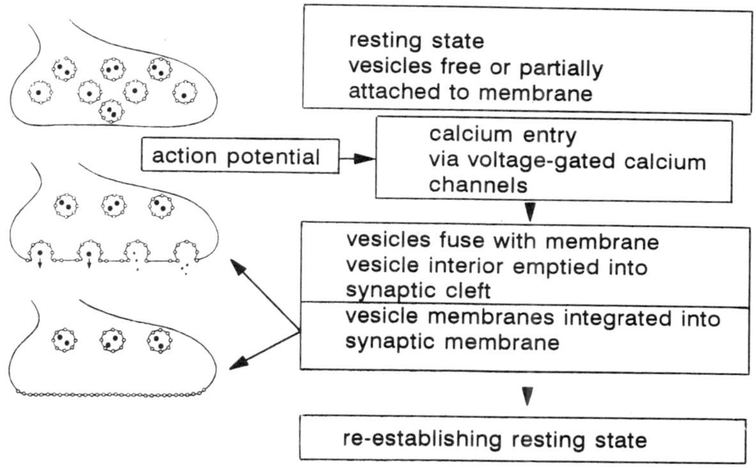


**Figure 6.5** Principles of synaptic transmission. The events at a chemical synapse (for example, the cholinergic synapse) are used to describe the entire process. At the right side of the Figure is a list of what is known about some important steps involved in the signal transmission in cholinergic synapses.

The fusion of vesicles with membranes is a process of great importance in biology, too. Exocytosis is simply the fusion of vesicles with membranes. In a very important example, we will shortly focus on signal transduction in chemical synapses, in which the fusion of so-called synaptic vesicles with the presynaptic membrane is a key step (Darnell *et al.*, 1986). Here the synaptic vesicles contain the transmitter in their inner volume and this transmitter is set free in the synaptic cleft by the fusion event. In Figure 6.5 the principles (important in membrane fusion) of synaptic transmission are summarized. For fusing protein-containing vesicles with pre-formed bilayers (artficial fusion) we can learn a lot from this naturally-occurring fusion process. We will not go into details about synaptic transduction here, the interested reader should refer to the literature; we just want to point out a few important principles, which are present in synaptic fusion as well as in bilayer reconstitution experiments.

The vesicles and the membrane coexist in a system and, as a first step, the vesicles must attach to the membrane. After this there must be a control mechanism to start the fusion process itself. By the fusion, the entire membrane content of the vesicles becomes an integral part of the postsynaptic membrane and (especially important in the case of synaptic vesicle fusion) the inner content of the vesicle is emptied into the synaptic cleft. After a certain number of fusion events the fusion must be stopped by another control mechanism. After that the system can be re-established and later a new cycle can be started.

Independent of the existence of control mechanisms, the fusion-process is itself a stochastic process which even now is not very well understood. The stochastic nature of membrane fusion can be very well monitored in nature by the existence of so called miniature endplate potentials which are the result of spontaneous fusion events of synaptic vesicles with the presynaptic membrane. By use of the patch-clamp technique a method has been created to measure exocytosis in cells (Neher and Marty, 1982). This method uses the fact that each time a vesicle fuses with the membrane its area is increased by the area of the vesicle membrane. Thus the increase in area can be measured as an increase in capacity of the cell membrane (which has a specific capacity of some 1 $\mu F/cm^2$). A diagram of the technique is given in Figure 6.6 together with a sketch of an experimental recording. In principle the same technique can be used in bilayer experiments to monitor vesicle fusion (see Chapter 3 for more details). Another technique to monitor fusion has already been presented in Figure 3.19. It makes use of another finding we already know from exocytosis; the interior volume of the vesicle can be emptied to the other side of the membrane (Zimmerberg *et al.*, 1980). What can be done is to load the vesicles with a dye before doing a fusion experiment and then add them to the aqueous

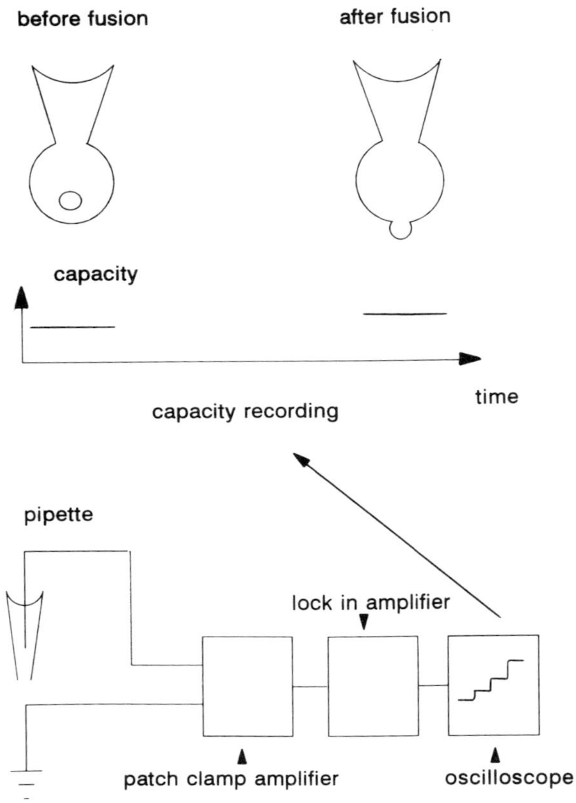


**Figure 6.6** Method to measure the electrical capacity of the membrane of a whole cell. The method is used to monitor fusion of vesicles during exocytosis via the increase in the area of the cell membrane after each fusion event. (More details of capacity measurement at biological membranes have been described in Chapter 3).

phase on only one side of the bilayer. When we then let them fuse with the bilayer and trace optically the dye on the other side of the bilayer, we can monitor the occurrence of dye on the other side of the bilayer as an indicator of vesicle fusion.

What we learn from nature when planning fusion experiments is that after making a bilayer and adding vesicles to the system (here the bath solution) we have to think about attaching the vesicles to the membrane, controlling the fusion process itself and handling the stochastic nature of the underlying biophysical process. All these steps are outlined later in this chapter.

Before going into more details of the fusion of vesicles with membranes in bilayer reconstitution experiments, we want to point to another line of experiments which has contributed a great deal to the understanding of membrane fusion: the fusion of vesicles with each other in vesicle solutions (Papahadjopoulos *et al.*, 1982). In these experiments mechanisms have been studied mainly to control the rate of fusion under different experimental conditions. As the principles are identical to bilayer fusion experiments all the data obtained can be used to set up bilayer reconstitution techniques. It was found that fusion rates under physiological conditions are usually very low, so the most important aim of controlling fusion must be to find conditions under which fusion rates are increased. The following list summarizes the results on how to improve fusion, collected from the enormous amount of literature available on this subject (Cohen *et al.*, 1980; Zimmerberg *et al.*, 1980; Boheim *et al.*, 1981; Miller 1986; Hanke and Breer, 1989; Neumann *et al.*, 1989; Niles and Cohen, 1991). Fusion is enhanced:

1. with membranes containing negatively charged lipids in the presence of calcium (or positively charged polypeptides) in the bathing solution
2. with membranes containing large amounts of kephalin (phosphatidylethanolamine)
3. by the presence of fusion enhancing drugs or viruses
4. by the presence of organic solvent in the membranes
5. with membranes made from *one* lipid component in the range of the temperature of the phase transitions of this lipid
6. by high electric fields (the fusion of cells and liposomes in high electric fields is itself an area of research.

All these points deal either with membrane composition or with membrane structure, this holds for point 3, too, as fusion-enhancing drugs most probably become an integral part of the membranes involved. One general interpretation of the above findings is that each of the points mentioned creates instabilities in the membrane, and these instabilities are obviously necessary to increase fusion rates. A diagram of the function of such a 'fusion-spot' is given in Figure 6.7. Before proceeding we want to add a few additional remarks to each of the points already given.

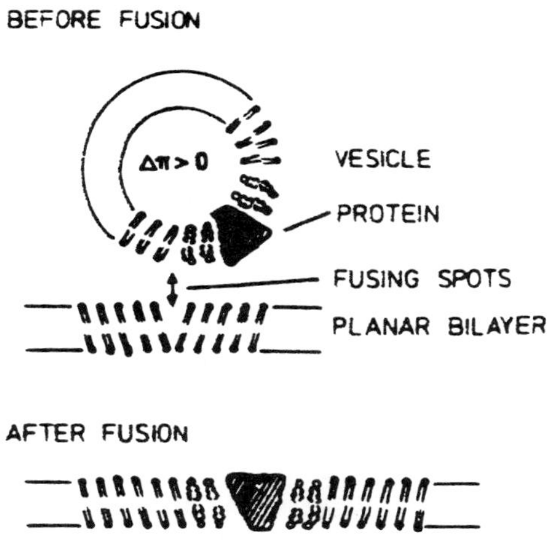


**Figure 6.7** Diagram of fusion spots which might be necessary to induce fusion of two membranes. In the lower part of the Figure a possible molecular mechanism of fusion is shown (Reproduced from Boheim *et al.*, 1982, with permission).

1. *Negatively charged lipids*. Natural membranes always contain a certain amount of negatively charged lipids. Typical negatively charged lipids are phosphatidylserine and phosphatidylinositol. Asolectin, for example, can contain about 10% phosphatidylinositol. Making a bilayer, the amount of negatively charged lipids can be varied from 0% to 100%. In our discussion 0% is of no relevance; on the other hand, at very high concentrations of negatively charged lipids, membranes tend to become unstable; furthermore, biological membranes are never made from negatively charged lipids only. When using divalent cations together with negatively charged lipids, 1 mM $CaCl_2$ usually is sufficient for inducing fusion. Instead of $Ca^{2+}$, which in our view is the best choice as cells use calcium for controlling exocytosis, positively charged polypeptides have been used successfully. Whereas $Ca^{2+}$ affects not only fusion but also ion-channel properties, and thus must be removed from the set-up in some experiments, this may not hold for positively charged polypeptides.
2. *Phosphatidylethanolamine (kephalin)* is known to induce non-bilayer (hexagonal) structures in planar lipid membranes. It is an ingredient of many biological membranes (*E. coli*, for example, has high amounts of kephalin in its membrane). Planar lipid bilayers can not usually be made from pure kephalin but should preferably contain only some 10%.
3. *Viruses and drugs*. It is known that some viral proteins enhance fusion. Phospholipase C (Niles and Cohen, 1991) and unsaturated lipid (Wolfe *et al.*, 1991) have also been described to enhance fusion. The same holds for nystatin (Woodbury and Miller, 1990) and annexin (Oshry *et al.*, 1991). Nystatin additionally forms pores in membranes. Each of these substances can be used, but one should bear in mind that another component which has to be studied before starting the planned experiments, is added to the system.
4. *Organic solvents*. This point is partially identical to point 3 and mentioned separately only because some organic solvent is present in nearly every planar lipid bilayer (see Chapter 5). For this reason, in critical applications (according to the fusion rate) painted bilayers obviously have some advantages independent of other factors.
5. *Phase transition*. The role of phase transitions in biological membranes is not very well understood. In any case it is impossible to define phase transitions in mixed systems, therefore a bilayer made from one well-defined lipid is a prerequisite of this technique. The phase transition of this lipid should be at an acceptable temperature, 10–35°C. Some synthetic lipids fulfilling these points have been described in the literature, dimyristoylphosphatidylcholine is one of them. When using this method the effect of the phase transition on the transport system under investigation must be known. A real disadvantage of the technique is that bilayers are mechanically very unstable in the range of the phase-transition temperature, thus not many experiments using this technique have been described in the literature. On the other hand, the fusion of vesicles in suspension has been very well investigated under conditions of lipid phase transitions, and some practical procedures, such as the freeze-thaw method to increase vesicle size make use of it.
6. *Electrical fields*. This point is mentioned

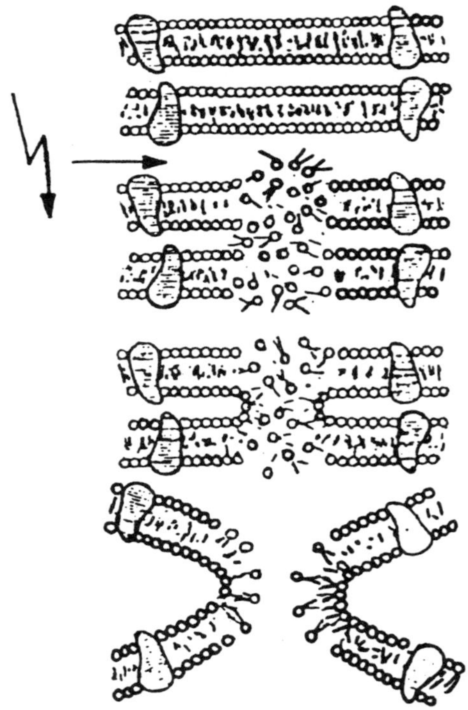

**Figure 6.8** Mechanistic steps involved in the fusion of two pure lipid membranes when a high electric field (pulse indicated by the arrow) is applied. First aqueous pores are made in the bilayers. When they are rearranged, the two bilayers may become one continuum.

mainly for reasons of completeness, it may be used, however, together with dip-stick membranes. Furthermore, the mechanistic steps involved in the fusion process itself are best understood together with electrofusion. A diagram of the fusion of two pure lipid membranes after a high electric field pulse is given in Figure 6.8. First a lipid pore is created and by rearrangement of the membranes fusion is achieved. Up to a certain degree this interpretation is an analogue of the idea of fusion spots.

Proceeding from vesicle–vesicle fusion to bilayer experiments, additional conditions have been found which are able to enhance fusion:

1. vigorous stirring of the bath solution
2. the vesicle interior should be hyperosmotic compared with the bath solution (at both sides of the bilayer)
3. high vesicle concentration
4. high bilayer area.

Points 3 and 4 are trivial and hold for any reconstitution technique. Point 1 invokes a technical problem; stirring may destroy the bilayer and must be stopped when starting the measurements.

### *6.2.2.2 Procedure*

With all this information we can now start to plan the first fusion experiment. At the outset one has to decide which bilayer technique is to be used. Using painted bilayers has the advantage of increasing the fusion rates by the presence of decane in the bilayer, but it also has the disadvantage of the presence of high concentration of organic solvent. Next, one has to think about the lipid composition of the bilayer; here points 1, 2 and 5 of the list in Section 6.2.2.1 are relevant. Having done this, one can start the experiment. Finally, the preparation to be reconstituted must be known. We give here a simple procedure for an initial fusion experiment:

1. Make a painted bilayer from Asolectin as described in the previous chapter. The bilayer will contain decane and negatively charged lipids, both increase fusion rates.
2. Use bath solutions containing 1 mM $CaCl_2$, 200 mM KCl and 10 mM Tris–HCl adjusted to pH = 7.4, thus sufficient $Ca^{2+}$ ions are present in the bathing solution.
3. Test the bilayer for stability and spontaneous fluctuations, only proceed if the bilayer is really stable.
4. Make a vesicle preparation (crude membrane preparation) from *Torpedo* electroplax as described in Chapter 4. These vesicles should be made in 500 mM KCl solution thus being hyperosmotic with the bath solutions (or use any other appropriate preparation).
5. Add 50 μl of the proteoliposome solution to the front side of the bilayer chamber (the *cis*-side).
6. Stir this side vigorously (see Section 3.1.7 for equipment).
7. Monitor current across the bilayer at a fixed potential (50 mV).

8. On the occurrence of channel-like current fluctuations stop stirring and add 2 mM EGTA to the vesicle side to prevent further fusion events.
9. Alternatively stop stirring and perfuse vesicle side with calcium- and vesicle free-solution (again to prevent further fusion).

When doing such experiments it was reproducibly found that each fusion event is accompanied by a so-called fusion spike, a transient increase in bilayer conductance. These fusion spikes can be used to count the total number of fusion events in an experiment, and they are also a useful tool to monitor fusion (point 7), an example of a real experiment is given in Figure 6.9.

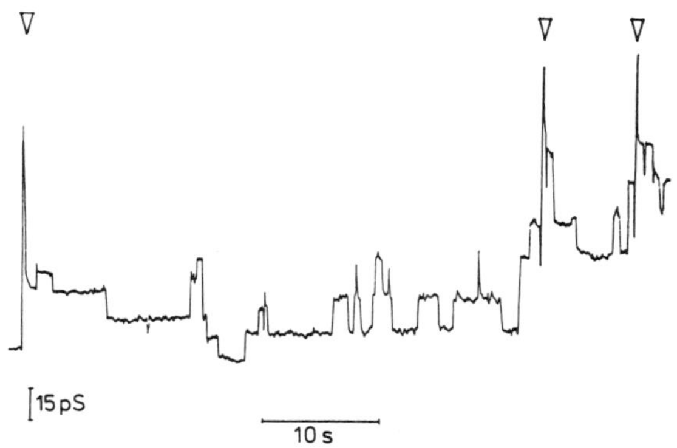


**Figure 6.9** Occurrence of fusion spikes in a bilayer reconstitution experiment. The transient spikes are marked by arrows. After each spike the bilayer conductance is increased as additional ion channels are incorporated, as can be seen by the single-channel fluctuations. Calcium-permeable channels of cilia from *Paramecium* were reconstituted in the experiment shown.

One of the advantages of bilayer reconstitution experiments is the possibility of experimenting with reconstituted ion channels at the real single-channel level. This can best be achieved by using a preparation of vesicles with very low protein density. Statistically when each liposome has only one or no ion channel in its membrane (or none), just one (or none) channel can be incorporated into a bilayer by a fusion event. When fusion is monitored by the occurrence of fusion spikes, according to points 8 and 9, fusion has to be stopped just after the first channel event shows up.

Of course the procedure given here is only a series of hints on how to do reconstitution experiments. As can be seen from the list of parameters affecting fusion rates, most of the points can be modified according to the needs of the experiment.

## 6.3. PIPETTE TECHNIQUES

All of the ideas given for planar lipid bilayers for incorporating proteins hold for dip-stick bilayers too. There are, however, a few points which need special attention in these techniques.

1. The area of a pipette bilayer is very small (a few square microns), thus the incorporation rates are low.
2. The pipette glass is hydrophilic and vesicles may attach to the glass.
3. Organic solvents can not be used to increase fusion rates (thus painted painted bilayers are not possible).
4. Stirring of the bath solution usually is not possible for mechanical reasons.

One has to bear all these points in mind when doing dip-stick experiments as they usually reduce the rate of success.

When using substances dissolved in water, there are no other differences from classical bilayer experiments.

It is tempting, however, to use the dip-stick technique with monolayers spread from protein containing liposomes. A diagram of the technique is given in Figure 6.10. We already have described all the general needs for this technique. Due to the ease of handling the dip-stick technique it is a good tool for fast screening of unknown biochemical preparations. Due to the small bilayer area it is useful to start with higher protein concentration than in classical bilayer experiments.

When trying to fuse liposomes with dip-stick bilayers, the liposomes can be added to the bath solution; there are no other differences from classical methods. When the vesicles are instead added to the pipette interior (Figure 6.11) it is often found that the fusion rates were higher than

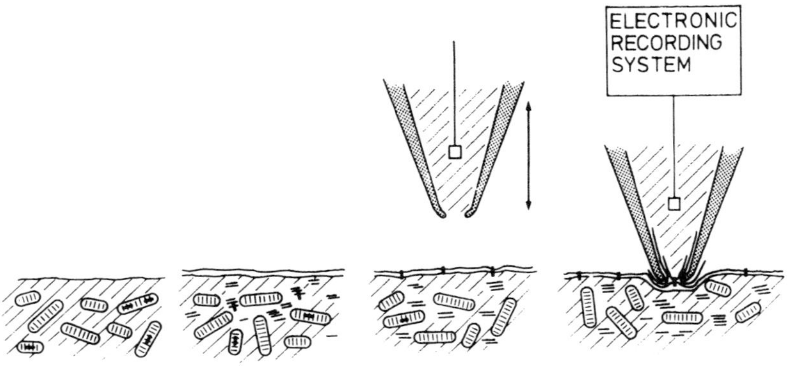


**Figure 6.10** The dip-stick technique used together with monolayers spread from protein-containing liposomes to reconstitute ion channels into planar lipid bilayers. Principles of the technique are identical with those of the technique of folded bilayers as shown in Figure 5.6.

expected from the small membrane area. This may be due to the fact that not only is the real bilayer at the tip of the pipette the active area for fusion, but also the total glass interior of the pipette is covered by a bilayer and active in fusion, allowing proteins to diffuse to the tip of the pipette after fusion of a liposome at another place. This statement is at least partially valid for all dip-stick reconstitution experiments.

All the reconstitution techniques have been used successfully with dip-stick bilayers. However, the reproducibility of classical techniques has up to now been found to be better than that of the dip-stick techniques, so we would recommend the dip-stick technique for fast screening investigations and for some special applications where, for example, high current and time resolution are needed in electrical measurements. In general, we think that the classical bilayer techniques are better for protein reconstitution into planar lipid bilayers.

For anyone having a patch-clamp set-up to hand this may be the main reason to use the dip-stick technique for reconstitution experiments. With such a set-up and dip-stick bilayers fast perfusion experiments, for example, can be done with purified proteins, comparable to what has been done with ion channels in patch-clamp experiments.

## 6.4 ATTACHMENT OF MEMBRANE FRAGMENTS TO PRE-FORMED BILAYERS

The incorporation of transport-mediating proteins into planar lipid membranes is one approach to making them accessible to electrophysiological experiments. However, especially when the

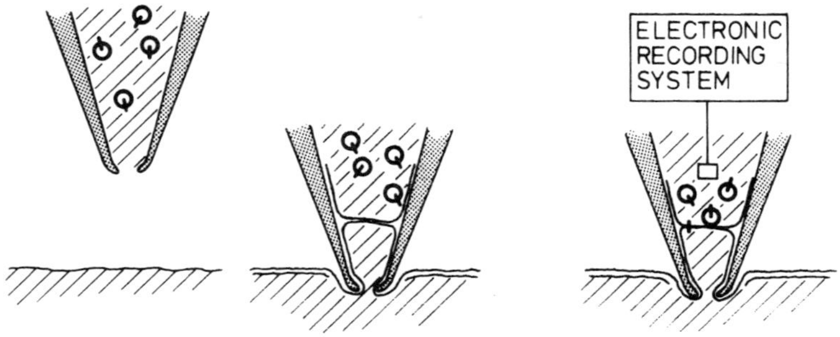


**Figure 6.11** Vesicles being fused with a dip-stick bilayer from the the pipette interior.

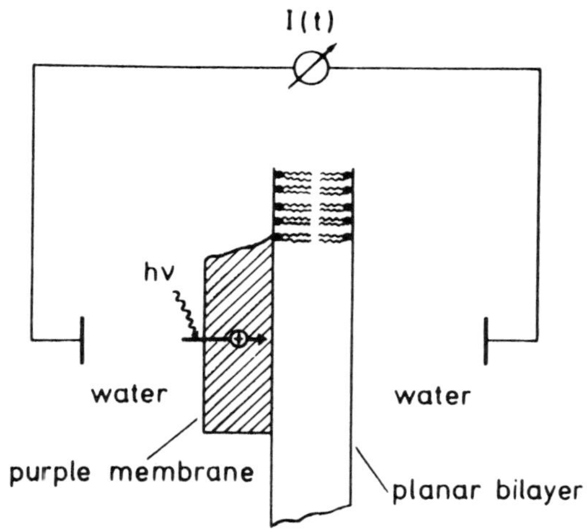


**Figure 6.12** Attachment of a membrane fragment to a pre-formed planar lipid bilayer. The system has a sandwich-like structure. Transport proteins in the attached fragment will modify the electrical properties of the whole system.

properties of ion carriers and pumps are investigated, it is not always the method of first choice. Bamberg and others (Fahr *et al.*, 1981) have developed another approach which sometimes is advantageous here. They have attached protein-containing membrane fragments to a pre-formed bilayer, thus creating a sandwich-like double-membrane system. An outline of the technique is shown in Figure 6.12, in this case using fragments from purple membranes in which bacteriorhodopsin is contained. The photo-current after flashing the system with a laser has been measured by use of this set-up. When applying the appropriate mathematics to such a system, it can be used for fast study of properties of transport proteins in the membrane fragment. We do not want to focus on this technique in detail here; nevertheless in the following a short outline of the technical steps involved is given.

Usually fragments from natural membranes are used, therefore negatively charged lipids are present in these fragments. To make them attach better to the bilayer it can be made with a small amount of positively charged lipids. As such lipids are not always available, an alternative is that bilayers should at least be made from neutral lipids or should be used together with divalent cations in the bath solution to minimize electrostatic effects. The remaining comments to be made are:

1. Any bilayer technique can be used, as the bilayer itself is not the object of interest; painted bilayers of moderate area are easiest to use.
2. Aqueous solutions have to be composed as necessary for the investigation of the transport system.
3. A small amount of membrane-fragment preparation is added to one side of the bilayer.
4. By monitoring the capacity and the conductance of the bilayer the spontaneous attachment of membrane fragments to the bilayer can be traced.
5. After a sufficient time of stirring (0.5–1 hr) fragments will be attached to the bilayer and the electrical measurements can begin.
6. Of course membrane fusion must be prevented in this type of experiment or the experiments must be stopped after fusion.
7. Sometimes the bilayer conductance in the experiments is increased by adding ionophores to it.

A simplified diagram of the electrical circuitry of such a system is given in Figure 6.13 together with a sketch of a typical experimental recording

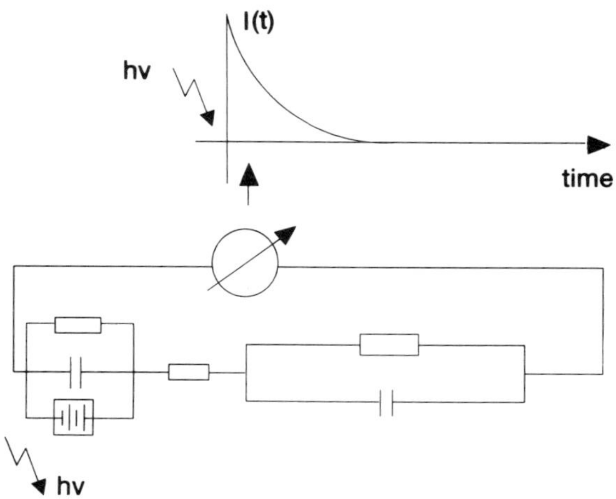


**Figure 6.13** An equivalent electrical circuit (simplified) of the system from Figure 6.12 is shown. When all the parameters of this electrical circuit are known, the conductance of the attached membrane and its time dependence can be measured as outlined.

(current recording after a bright flash). More details of this technique are given in the literature.

## REFERENCES


Anderson, O.S. and Koeppe, R.E. (1992) Molecular determinants of channel function. *Physiol. Rev.* **72**, 89–158.

Benz, R. and Läuger, P. (1976) Kinetic analysis of carrier-mediated ion transport by the charge-pulse technique. *J. Membr. Biol.* **27**, 178–196.

Benz, R., Schmid, A., Nakae, T. and Vos-Scheperkeuter, G.H. (1986) Pore formation by LamB of *Escherichia coli* in lipid bilayer membranes. *J. Bacteriol.* **165**, 978–986.

Boheim, G. (1974) Statistical analysis of alamethicin channels in black lipid membranes. *J. Membr. Biol.* **19**, 277–303.

Boheim, G., Hanke, W., Methfessel, C., Eibl, H., Kaupp, U.B., Maelicke, A. and Schultz, E. (1982) Membrane reconstitution below lipid phase transition. In: *Transport in biomembranes: Model systems and reconstitution*. R. Antolini *et al.*, (eds.), Raven Press, New York, pp. 87–97.

Cohen, J.A., Zimmerberg, J. and Finkelstein, A. (1980) Fusion of phospholipid vesicles with planar phospholipid bilayer membranes: II. Incorporation of a vesicular marker into the planar membrane. *J. Gen. Physiol.* **75**, 251–270.

Darnell, J., Lodish, H. and Baltimore, D. (1986) Molecular cell biology. Scientific American Books, New York.

Darzon, A. (1986) Planar bilayers: A powerful tool to study membrane proteins involved in transport. In: Methods in enzymology Vol. 127. C. Packer, (ed.), Academic Press, Orlando, Florida, pp. 486–502.

Fahr, A., Läuger, P. and Bamberg, E. (1981) Photocurrent kinetics of purple-membrane sheets bound to planar lipid membranes. *J. Membr. Biol.* **60**, 51–62.

Hanke, W. (1985) Reconstitution of ion channels. *CRC Crit. Rev. Biochem.* **19**, 1–44.

Hanke, W. and Breer, H. (1986) Characterization of the channel properties of a neuronal acetylcholine receptor reconstituted into planar lipid bilayers. *J. Gen. Physiol.* **90**, 855–879.

Hanke, W. and Breer, H. (1989) Reconstitution of acetylcholine receptors into planar lipid bilayers. In *Subcellular Biochemistry*, Vol. 14, J.R. Harris and A.H. Etmadi (eds.), Plenum Press, New York.

Hladky, S.B. and Haydon, D.A. (1984) Ion movements in gramicidin channels. *Curr. Top. Membr. Transp.* **21**, 327–372.

Latorre, R. (1987) *Ionic channels in cells and model systems*. Plenum Press, New York.

Miller, C. (1983) Integral membrane channels: Studies in model membranes. *Physiol. Rev.* **63**, 1209–1242.

Miller, C. (1986) *Ion channel reconstitution*. Plenum Press, New York.

Montal, M., Anholt, R. and LaBarca, P. (1986) The reconstituted acetylcholine receptor. In *Ion channel reconstitution*. C. Miller, (ed.). Plenum Press, New York, USA, pp. 157–196.

Neher, E. and Marty, A.(1982) Discrete changes of cell membrane capacity observed under conditions of enhanced secretion in bovine adrenal chromaffin cells. *Proc. Natl. Acad. Sci. USA* **79**, 6712–6716.

Neumann, E, Sowers, A.E. and Jordan, C.A. (1989) *Electroporation and electrofusion in cell biology*. Plenum Press, New York.

Niles, W.D. and Cohen, F.S. (1991) Fusion of influenza virions with a planar lipid membrane detected by video fluorescence microscopy. *J. Gen. Physiol.* **97**, 1101–1119.

Oshry, L., Meers, P., Mealy, T. and Tauber, A.I. (1991) Annexin-mediated membrane fusion of human neutrophil plasma membranes and phospholipid vesicles. *Biochem. Biophys. Acta* **1066**, 239–244.

Papahadjopoulos, D., Poste, G. and Vail, W.J. (1982) In: *Methods in membrane biology*, Vol. 10, E.D. Korn (ed.), Plenum Press, New York, pp. 1–122.

Schindler, H. (1982) Concept and techniques for membrane transport reconstitution. In: *Model systems and reconstitution*, R. Antolini *et al.*, (eds.), Raven Press, New York, pp. 75–85.

Schindler, H. (1989) Planar lipid protein membranes: Strategies of formation and of detecting dependencies of ion transport function on membrane conditions. *Methods Enzymol.* **171**, 225–253.

Tamm, L.K. (1991) Membrane insertion and lateral mobility of synthetic amphiphilic signal peptides in lipid model membranes. *Biochim. Biophys. Acta* **1071**, 123–148.

Ti Tien, H. (1974) Bilayer lipid membranes: Theory and application. Marcel Dekker Inc., New York.

Wolfe, J., Prez, E., Bonanno, M. and Chapel, J.-P. (1991) The interaction and fusion bilayers formed from unsaturated lipids. *Eur. Biophys. J.* **19**, 275–281.

Woolley, G.A. and Wallace, B.A. (1992) Model ion channels: Gramicidin and alamethicin. *J. Membr. Biol.* **129**, 109–136.

Zimmerberg, J., Cohen, F.S. and Finkelstein, A. (1980) Fusion of phospholipid vesicles with planar lipid bilayer membranes: I. Discharge of vesicular contents across the planar membrane. *J. Gen. Physiol.* **75**, 241–250.

CHAPTER SEVEN

# Data acquisition and evaluation

While performing a series of successful bilayer experiments, data will accrue rapidly and must be collected, interpreted, and either saved for later evaluation, or directly evaluated during the experiments. In this chapter we will make a few remarks on data structures typical of bilayer experiments. In addition to the comments made in this chapter, we want to refer the reader to the fact that most of the data evaluation procedures used for patch-clamp experiments can be used for bilayer experiments too. This holds especially when single-channel data have to be handled (Sakmann and Neher, 1983).

## 7.1 STRUCTURE AND TYPES OF DATA OBTAINED

The data from bilayer experiments can be roughly classified as given in the following sections. Of course additional points are possible, but here we are focusing on the most typical and important data types and questions concerning how to handle them.

### 7.1.1 Electrical properties of pure lipid membranes

Data on the electrical properties of pure lipid membranes are the data most typical of bilayer experiments. They are obtained exclusively from bilayer experiments (or liposomes), as natural membranes are never pure lipid membranes.

Typically the capacity and the resistance (or conductance) of bilayers are used to describe their basic electrical properties. These can be measured by equipment as described in Chapter 3, using bilayers made according to any of the techniques described in Chapter 5. Additionally, these parameters are frequently used to control the quality of the bilayer before starting, for example, a reconstitution experiment as described in Chapters 6 and 8. In certain groups of experiments, however, the basic electrical parameters

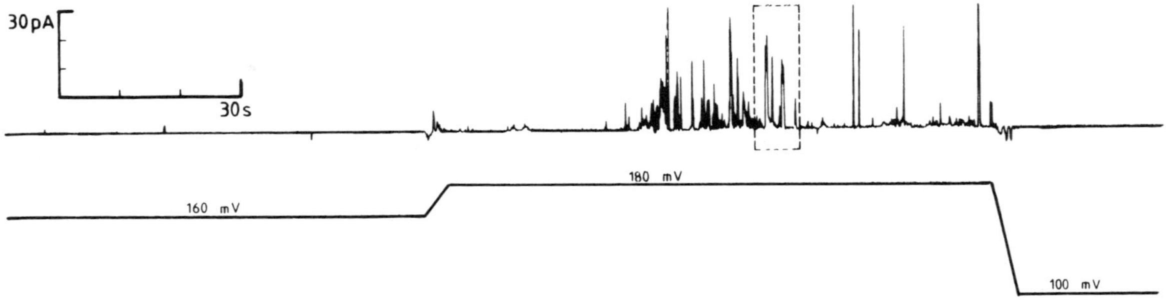


**Figure 7.1** Results of a voltage jump experiment, clamping a pure lipid bilayer to a comparatively high potential (180 mV). After the pulse is switched on (starting from a lower potential), current fluctations start and can then be observed as long as the potential is high, here in a lipid membrane from one synthetic lipid. When the potential is switched down to a lower value, the current fluctuations disappear after a short delay. For convenience the applied potential and the measured current are shown as functions of time. The data were kindly provided by J. Maack (Maack, 1991).

of pure lipid membranes are measured as function of physical parameters, such as temperature and pressure, or as a function of chemical parameters such as the pH of the aqueous solution or the chain length of the lipid used.

Electrical parameters of bilayers are usually acquired as values read from the bilayer amplifier (current, applied voltage, capacity) and can be stored (typically on a tape recorder or on a computer, as described in Chapter 3) for later use. For example, the capacity and conductivity of a bilayer made from a pure lipid with a phase transition at room temperature, can be plotted as function of temperature (Boheim et al, 1980). A detailed descripition of such an experiment is given in Section 8.1.1.

Another group of experiments typically handling electrical data on pure lipid membranes is the investigation of bilayer behaviour at high potential. Here it has been shown that pure lipid membranes tend to form aqueous pores when short pulses of high potential are applied. Comparable effects can be observed when an intermediate potential (200 mV) is applied to a bilayer for a longer time, some seconds (Maack, 1991). Such an experiment is described in Figure 7.1 in some detail. Pore formation in pure lipid membranes at high potential can be studied and used in so-called electroporation experiments.

### 7.1.2 Ion carrier data

When ion carriers are incorporated into planar lipid bilayers the current across such a bilayer can be measured, depending on various conditions, such as:

1. temperature
2. composition of the aqueous solution
3. lipid composition
4. potential.

The current measured in such experiments as a function of one of the above parameters can be used directly for the interpretation of the results. Sometimes, however, additional procedures have to be applied to the recorded data. One of these procedures is noise analysis (see Kolb and Läuger, 1978). By this procedure, in which the current which is measured as a function of time (and of course other parameters), the power density is calculated as a function of frequency by a Fourier transformation. A computer is necessary to do this (see later). Without going into details noise analysis provides information about the transport capacity of the basic unit involved.

In bilayer experiments the real number of incorporated ion-carrier molecules usually can not be determined, but from a noise analysis, using certain assumptions, such data can be evaluated, together with data about the transport rate of the single carrier molecule.

In addition to experiments discussed so far, which in general are done at fixed membrane potentials, others can be done with stepwise changes in membrane potential. In such experiments the time-dependent response of the current following a stepwise potential change is recorded (Läuger *et al.*, 1981). Not only the response to stepwise changes in potential but also the current response to stepwise changes in other parameters can be measured and information about the kinetics of the carrier transport can be obtained. These experiments are often called clamp experiments (voltage-clamp, current-clamp, concentration-clamp, etc.).

### 7.1.3 Single-channel data

One of the most important groups of bilayer experiments is the reconstitution and investigation of ion-channel proteins (and other molecules) into artificial membranes. Most tempting is to incorporate just *one* functional ion-channel forming protein into a bilayer and then to measure its 'single-channel' properties. Again this can only very rarely be done with patch-clamp techniques as the channel density in natural membranes usually cannot be controlled. In any case, the part of the membrane which is hit in a patch-clamp experiment and the number of channels included in it, are parameters which can not clearly be predicted. When doing such a 'single-channel experiment', even in bilayers, verification that only *one* functional channel forming molecule has been incorporated is important. To obtain reliable information on this problem is an aim of the statistical data evaluation done with the results from such an experiment (Horn, 1991).

The current across the bilayer in single-channel experiments can be measured with a fixed set of experimental parameters as function of time. An idealized sketch of such a trace is given in Figure 7.2. It is made on the assumption that the channel has only one electrically distinct open state. Thus the channel can only be closed – no current across the bilayer measured at a fixed potential – or open – a well defined current is flowing. From such a very simple trace a few basic parameters to describe a single ion channel are obvious:

1. open-state conductance
2. open-state lifetime
3. closed-state lifetime
4. probability of the channel being in the open state.

Point 4 of this list is always related to points 2 and 3 by a strict mathematical relation (see later). As the opening and closing of an ion channel is a stochastic process, all the above data have to

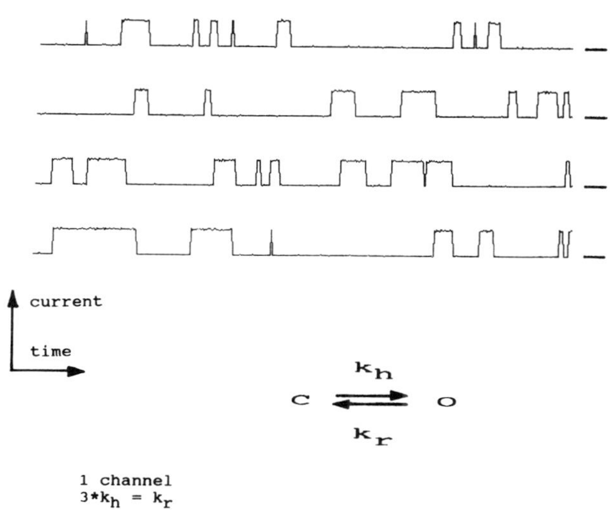


**Figure 7.2** Simulation of a single-channel fluctuation with only one open (o) and one closed (c) channel state, as given in the gating model below. The closed state is marked to the right of the traces, the current is plotted upward, and current and time are scaled arbitrarily. Transition rates between states are given by rate constants $k_h$ and $k_r$.

be calculated as averages over a larger number of events in real experiments.

Typically we have to make a few assumptions about the function of an ion channel when starting an analysis as described above. In principle these assumptions are also requirements of a functional ion channel in a natural membrane. A list of the more important assumptions follows:

1. Ion channels are gated, they exist in an open and a closed state (ion channels that are always closed usually can not be investigated, ion channels which are always open, only under very specific conditions are acceptable for a biological organism but will not be treated here).
2. Ion channels are permeable only to certain ions or molecules (they are selective).
3. Ion channel gating is controlled by relevant biological parameters (again exceptions exist, but these are not discussed in this chapter).

From this list a few conclusions can be drawn according to the handling of single ion channel data. All of the parameters of an ion channel can be measured as function of one of the parameters controlling its single-channel parameters as partially listed above.

Of course the real world of bilayer experiments is much more complicated than described in this first oversimplified introduction. Typically, an ion channel may have more than one electrical distinct open state and may follow an extremely complicated gating scheme. By gating scheme we mean the sum of all states an ion channel can adopt, the transitions between these states and the dependence of these states and transitions on all known (and unknown) experimental parameters. Such a scheme will completely describe the behaviour of an ion channel. Details about determining, for example, the gating scheme of an ion channel from a given set of experimental data are found in the original papers (Colquhoun and Hawkes, 1977, 1981, 1982; Hille, 1992).

Independent of these more complicated approaches, a few basic things must be done with results from a single channel experiment. By visual inspection of the current fluctuation traces, one can see whether channel fluctuations exist at all. One can find out whether one or more electrically distinct open states are present. It has to be determined whether one or eventually more than one channel has been incorporated into a bilayer. In the simplest situation this can be typically recognized when multiple events as shown in Figure 7.3 are present in a current recording. Having made such a rough first inspection, one can proceed to a more detailed data evaluation. Instead of describing such an evaluation here we will demonstrate in Section 8.2 and 8.3 a detailed data evaluation using alamethicin and the nAChR as examples.

### 7.1.4 Multi-channel data

Everything said for single channel data will hold multi channel data too. Additionally, one has

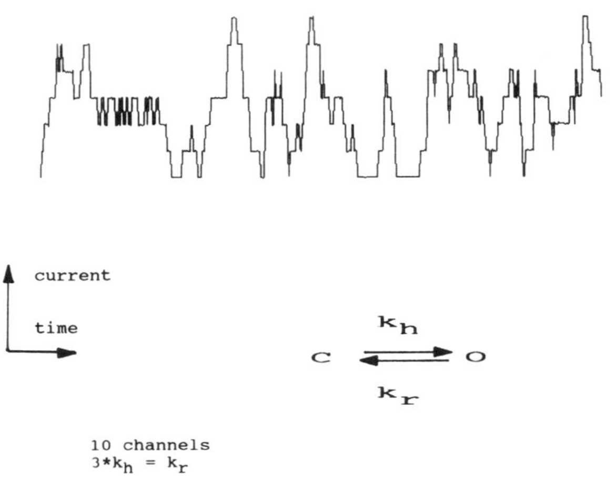


**Figure 7.3** A multi-channel fluctuation is simulated here, again with a channel type having only one open and one closed state. The closed state (no channel open) is marked to the right of the trace, the current is plotted upward, and current and time are scaled arbitrarily. In cases where there are different channels or channels with more than two electrical distinct states present in the bilayer, the situation becomes even less clear.

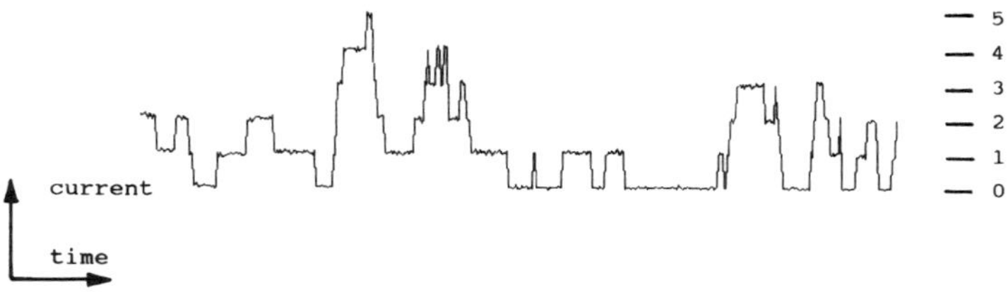


5 distinct open levels

all steps of equal distance

assuming 5 identical channels with an open state probability of $P_o$ and a closed state probability of $P_c$, the following holds:

$$P_c = 1 - P_o$$

5 channels open $P_{5o} = P_o^5$

all channels closed $P_{5c} = (1 - P_o)^5$

**Figure 7.4** Scheme for obtaining single-channel parameters from a multi-channel fluctuation. From the trace, the maximum number of open states is found. Then whether identical or different open states are involved is determined and finally some single parameters (amplitudes) are extracted.

to calculate the number of channels responsible for the measured results (Horn, 1991). The easiest but not very accurate procedure is to count the maximum number of channels open at the same time. Especially at a low channel open state probability this may lead, however, to inaccurate results. If the open state probability ($P_o$) of the channels can be controlled by an experimental parameter, it is better to choose conditions with a $P_o$ close to 1, as the statistical estimate about the number of channels involved in a given set of data will be more accurate under these conditions. When the number of involved channels has been thus identified (with a limited probability), further evaluation has to be done to extract the parameters of a single channel from the multi-channel data. As long as only a small number of channels is incorporated in a bilayer, this is possible. The principles of such an evaluation are summarized in Figure 7.4.

When a bigger number of channels is incorporated into a bilayer, other procedures have to be used, which are mainly identical to those used for ion-carrier data (see earlier). A problem complicating the situation is that in all such multi-channel data evaluations information about the open state probability of the single channel is required. Additionally, the assumption of identical but independent behaviour of all channels has to be made (this assumption has to be verified somehow: in cases of reconstituting a purified channel protein into a planar lipid bilayer, such an assumption seems to be acceptable). The information about open-state probability, however, can only be obtained from single-channel data. As a consequence of this dilemma, very often in multi-channel data analysis a $P_o$ being much smaller than 1 is assumed. As a result of an inaccurate assumption quite frequently the single channel parameters are calculated inaccurately from such experiments as can be found in some examples in the literature.

## 7.2 PRELIMINARY DATA EVALUATION BY EYE AND HAND

Starting with very simple experiments (and without sophisticated equipment) very often the current recordings from bilayer experiments will be obtained as strip chart recordings. Even from such strip charts much detailed information can be got by simple inspection and data evaluation. In Figure 7.5 we show such a recording from a single-channel experiment together with its analysis by hand and eye. The length and amplitude of each channel state is measured and added to a list. From this list arithmetic averages are calculated for state lifetimes and channel amplitudes. This is done for different sets of parameters and the results can be drawn as graphs.

Doing such a data analysis without computer is a good exercise to start with. After having

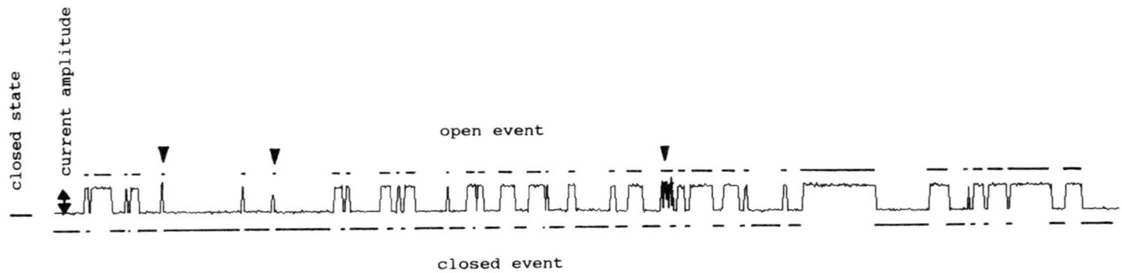


| No. | closed time (ms) | open time (ms) | amplitude (pA) | comment |
|---|---|---|---|---|
| 1 | | | | |
| 2 | | | | |
| 3 | | | | |
| 4 | | | | |
| 5 | | | | bad |
| 6 | | | | |
| 7 | | | | bad |
| 8 | | | | |
| 9 | | | | |
| 10 | | | | |

**Figure 7.5** Evaluating single-channel parameters by hand and eye can be done as shown here. The length and amplitude of each event in the trace is measured and added to a table for further evaluation, as shown in the lower part. 'Bad' and 'good' events are selected by eye (here marked by arrows).

measured the first let us say 10 000 single channel events as described (in case no computer is available, for example), everyone will be quite happy about any computer support and additionally will know the basic problems of such a data evaluation which are, among others:

1. limited time resolution
2. missing events
3. inaccuracy in amplitude and lifetime measurement
4. limitation in number of events to be evaluated.

Of course, procedures have been developed to correct for all these problems; indeed such mathematical tools must be used to obtain acceptable results.

After this first look at the difficulties of data evaluation we want to proceed to what usually is done in a bilayer laboratory today.

## 7.3 DATA EVALUATION USING COMPUTERS

Most probably, everyone will agree that today for efficient research it is no longer correct to evaluate experimental data by eye and by hand. Additionally many mathematical routines to statistically fit data with certain models are not possible without computer support. Thus in most electrophysiological and bilayer laboratories computers are used for data aquisition and data evaluation. We therefore here want to make a few principal remarks on doing so.

### 7.3.1 Technical requirements

Before proceeding to the real data analysis procedures using a computer, a few technical needs have to be fulfilled:

1. The data must be stored somewhere; usually magnetic tape recorders are used, due to the volume of data, which can not be stored elsewhere properly. With bigger mass-storage units available on a computer, data can be directly stored on the computer (via an A/D-converter).
2. The data must be digitized; this requires a proper analogue to digital converter which has a sufficient sampling rate (some kHz are necessary, see Chapter 3).

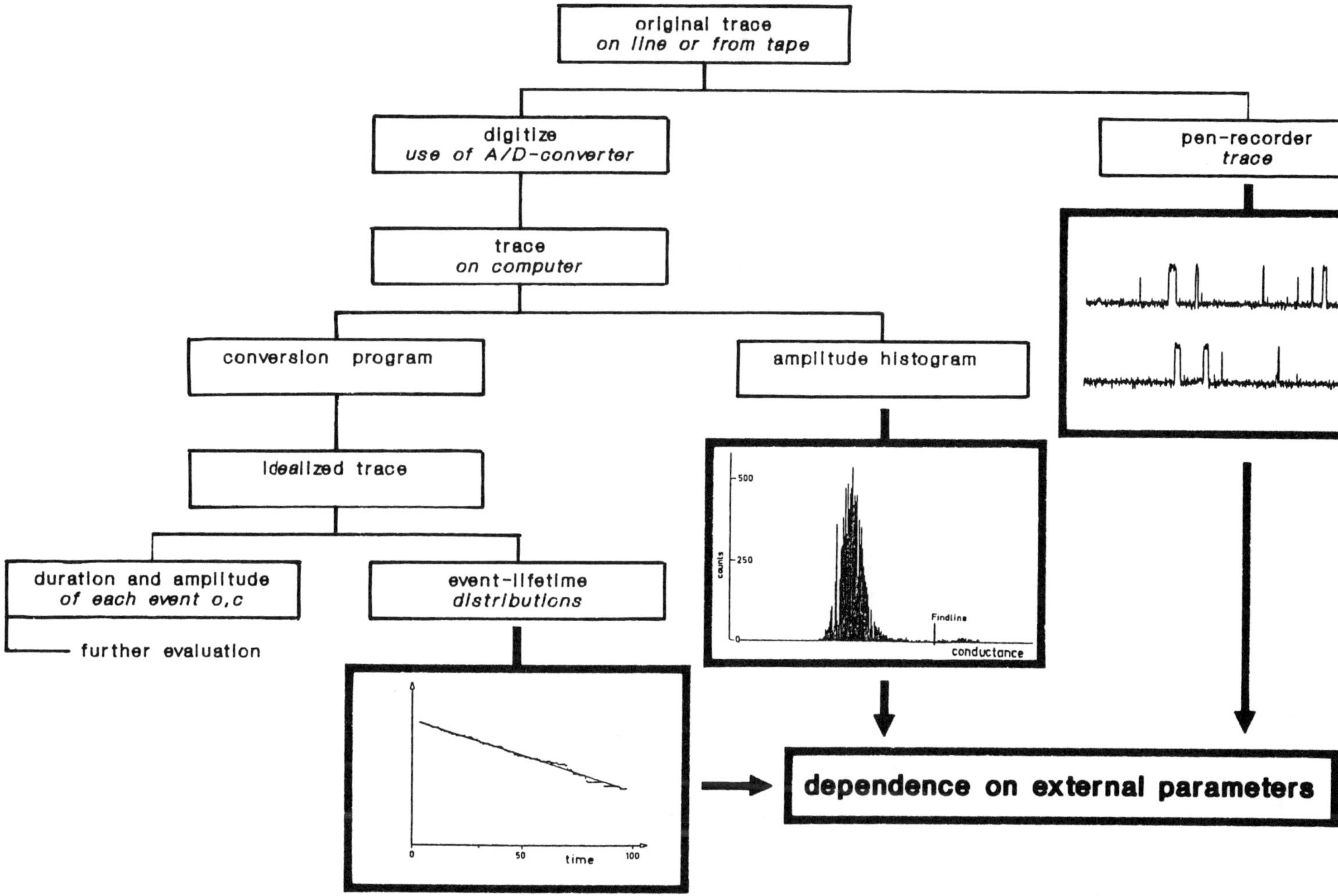


**Figure 7.6** Principles of a single-channel data evaluation as done with a computer (Reprinted and modified from Hanke and Breer, 1989, with permission).

3. Adequate software must be available for the acquisition and handling of the data.

All the above stated needs can be fulfilled by the use of commercially available equipment (see Appendix 4). We have already made some comments about computer hardware in Section 3.4. and thus can directly proceed to the questions about software needed in computer-supported data evaluation (see Sachs *et al.*, 1983).

### 7.3.2 Structure of software for data analysis

A typical data evaluation of single-channel data by a computer is shown in Figure 7.6. For the really important steps given in Figure 7.6, we will give some more detailed hints on how to proceed with the computer and on how to make it do the desired calculations. To make the situation easier, we have written a (simple) program in BASIC, which we want to use as an example when going through the analysis. A few details of this program are given in the following. As is usually done in developing a computer program first a flow chart is made which contains all the basic information about what the program has to do. An outline of such a flow chart is shown in Figure 7.7. After having created the flow chart, an overlay is programmed from which all functions of the program are accessible in 'pull-down menus'. In Figure 7.8, the start up screen and main menu of the program written by the authors is shown. From such a main-menu all other functions are accessible by the pull-down menus (some of these are given in Figure 7.8 too). In the following, we will go through some of the single procedures of this program as given by the different points of the menus.

The program starts from data which have been digitized by an A/D-converter and are stored on files on tape or disk. From these files, the data are imported into the program, for example from an ASCII file as shown in Figure 7.8 in one of the pull-down menus. The second step is a first inspection of the data on the screen (comparable to an oscilloscope). In Figure 7.9 such a screen is shown (this Figure is comparable to a hard-copy from a channel current fluctuation trace; different manipulations of such a trace are possible on the computer such as smoothing, zooming and others). Sometimes a direct hard-copy (by use of a printer) of such a screen is required for documentation. In a third step a total amplitude histogram of a channel fluctuation trace can be done to see how many electrical

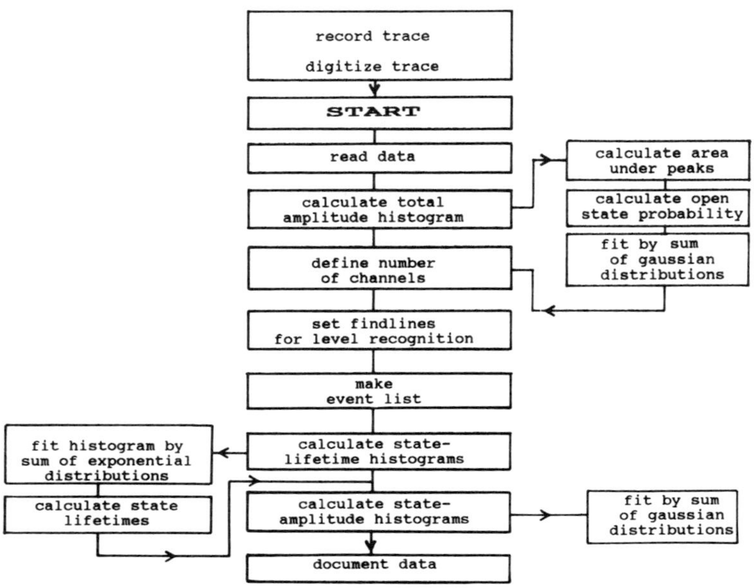


**Figure 7.7** Main steps that a program that does single-channel data evaluation by computer should have. A hard-copy should be possible at any stage of the data evaluation.

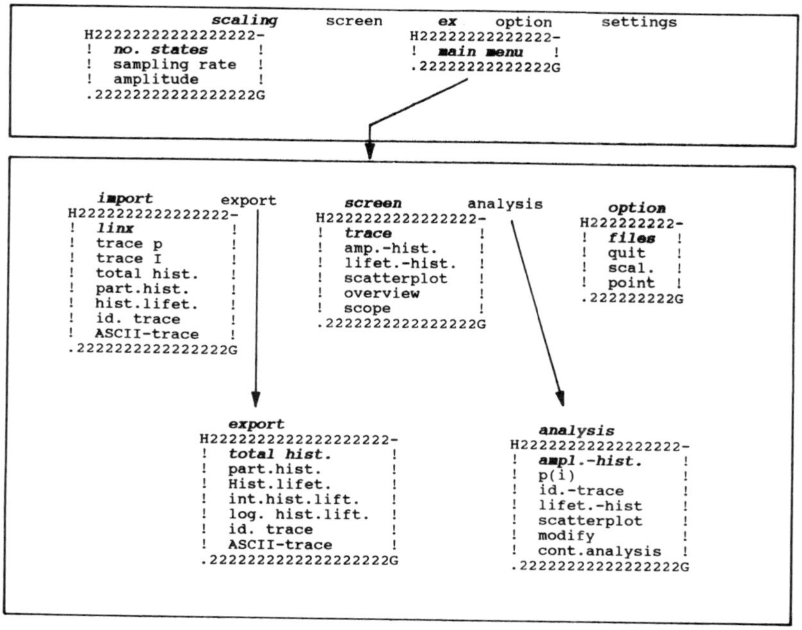


**Figure 7.8** Presentation of the main menu of a program written by the authors, together with all the pull-down menus. At some points in the primary pull-down menu, further menus will open. At the top, the start-up screen is shown, from which access to the main menu is made.

**Figure 7.9** Hard-copy of a computer screen using the scope function of the program to evaluate single-channel data (see Figures 7.6 to 7.8).

distinct states are found in the trace. From such an integral amplitude histogram it can, for example, be decided whether one or more channels are involved in the measured activity. An example is given in Figure 7.10. After having inspected the amplitude histogram and perhaps having printed another hard-copy it is time to proceed to what has already been done by eye and hand and what is most important of a computer analysis of single channel data: the recognition of each single event and the measurement of its duration and amplitude. To do this, certain criteria have to be given to the computer to enable the program to decide what a channel event is. The principle of setting findlines to define channel states is summarized in Figure 7.11. Typical additional criteria used for channel-state recognition are:

1. the amplitude range which is accepted for a certain channel state
2. the minimum duration which is stated to be necessary for a channel state
3. the stability of the amplitude during the event.

Other parameters may be necessary when the data are more complicated. When a list has been calculated with the duration and amplitude of each existing channel event, amplitude histograms and lifetime distributions can be created and fitted with certain functions, to exactly determine channel-state lifetime and amplitude, and then

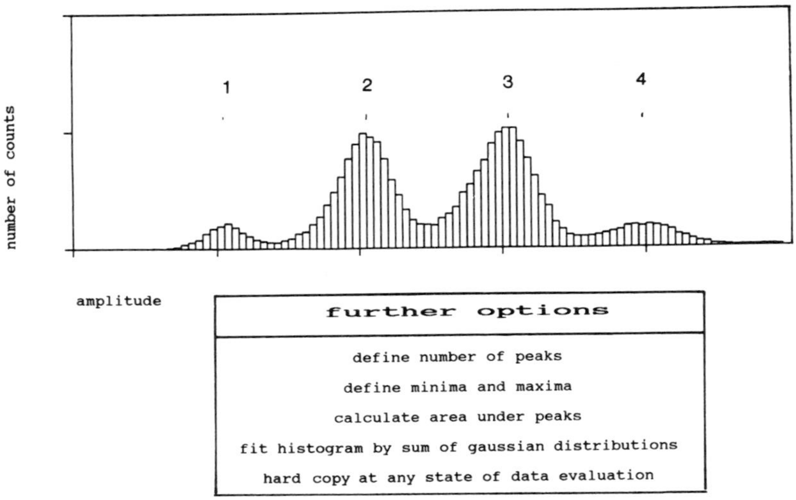


**Figure 7.10** Total amplitude histogram taken from a current fluctuation by a computer. Additional to the histogram, further options of the program (at this stage in the data evaluation) are shown.

to create graphs of these parameters as functions of experimental parameters.

At this point, a first (partially) complete analysis has been done and a basic set of data about a single channel fluctuation is present. Proceeding from here, the statistical evaluation can be taken to almost every state of complexity, thus we will stop here, referring the reader to the literature (Sakmann and Neher, 1983; Hille, 1992), and will proceed with some more general remarks.

What we have shown above using our own program can of course be done just as well (most probably better) with commercial software. Source codes of their programs are available from some commercial suppliers (sometimes for an additional fee), addresses are given in Appendix 4. It may be useful for those interested in computer aided data evaluation to use these source codes for better understanding. Furthermore, many programs are available for general data evaluation, graphics and statistics which can be used for evaluating and presenting data from bilayer experiments. Of course, general programs that exist for desk-top publishing and word-processing must be included in this list.

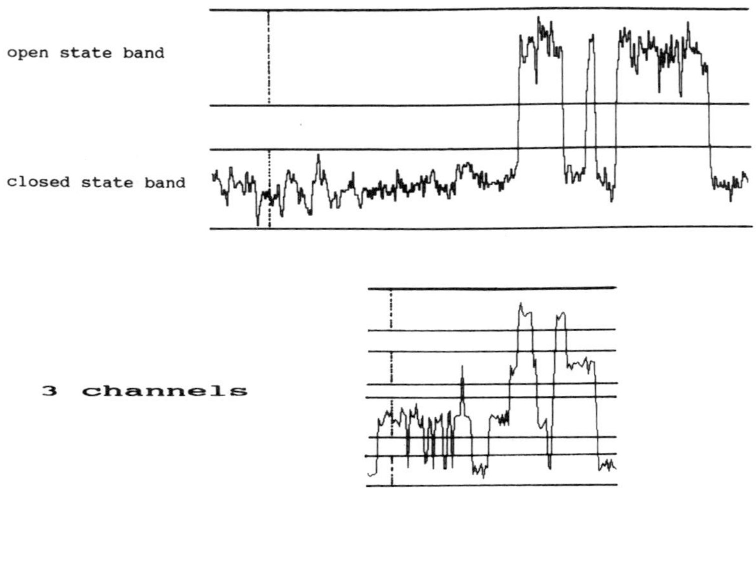


**Figure 7.11** Principles of single-channel recognition by a computer program. Findlines are set to define the different channel states and additional criteria are used to detect 'good' and 'bad' events. The procedure shown here can be used to evaluate multi-channel data as long as the different fluctuation states can be clearly resolved (see lower part).

## 7.4 GENERAL COMMENTS

In this chapter so far besides some more general statements, we have mainly focused on electrical data from bilayer experiments and even more on the evaluation of single-channel data, as this is the most typical case for bilayer experiments to date and also reflects our own interests. Many different experiments and hence analyses are possible with planar lipid bilayers, but comprehensive discussion of these is beyond the scope of this volume.

During the last few years, besides electrical measurements, another group of investigations and thus data has become common in the description and investigation of biological and artificial membranes. This is the use of optical methods, as was already mentioned a few times in earlier chapters. Some optical methods used in bilayer experiments are:

1. the investigation of the bilayer by direct optical inspection, by eye or by use of a video camera
2. the use of fluorescent dyes to measure the membrane potential (Haugland, 1992)
3. the use of dyes to measure ion concentrations in the bathing solutions (Haugland, 1992)
4. the use of dyes to measure the incorporation of external material in a bilayer (Niles and Cohen, 1991)
5. photobleaching experiments to measure mobility of molecules in a bilayer.

Of course this list is not complete. In the light of the importance of optical methods in bilayer research, we would like to risk the following statement here: within the coming years, the use of optical methods in all types of bilayer experiments will become almost as frequent as measuring electrical parameters is today. We will give a few examples of what can be done already:

1. *Direct optical inspection*. When measuring the black area of a coloured film (painted bilayer) with a video-camera and a video-imaging system as a function of time, information can be obtained about the diffusion of solvent out of such a film.
2. *Membrane potential*. The use of fluorescent dyes to measure membrane potential can substitute for classical electrical recordings in bilayer systems to determine, for example, the influence of charged molecules on the membrane potential.
3. *Ion concentration*. Dyes can be used to measure the transport of ions across a bilayer (see Section 6.2.2).
4. *Incorporation of material in bilayers*. Dyes can be used to measure the fusion of liposomes with a bilayer by tracing the incorporation of liposomal membranes into the bilayer.
5. *Photobleaching*. By bleaching an area of a bilayer which contains an appropriate dye, the diffusion of bilayer material from the unbleached region into the bleached region can be measured and used to calculate (lateral) diffusion coefficients.

Another statement is necessary at the end of this chapter. Independent of the type of data, highly sophisticated mathematical and statistical procedures to interpret data can be applied to learning everything about a certain transport system reconstituted into a planar lipid bilayer or the bilayer itself. We have outlined a variety of approaches which can be applied to pure lipid bilayers and to transport systems in previous chapters and will not go into further details.

## REFERENCES

Colquhoun, D. and Hawkes, A.G. (1977) Relaxation and fluctuations of membrane currents that flow through drug-operated ion-channels. *Proc. R. Soc. (Lond) [Biol.]* **199**, 231–262.

Colquhoun, D. and Hawkes, A.G. (1981) On the stochastic properties of single ion channels. *Proc, R. Soc. (Lond) [Biol.]* **211**, 205–235.

Colquhoun, D. and Hawkes, A.G. (1982) On the stochastic properties of bursts of single ion channel openings and of clusters of bursts. *Phil. Trans. R. Soc. (Lond) [Biol.]* **300**, 1–59.

Hanke, W. and Breer, H. (1989) Reconstitution of acetylcholine receptors into planar lipid bilayers. In: *Subcellular biochemistry*. J.R. Harris and A.H. Etmadi, (eds.), Plenum Publishing Corporation, New York, pp. 339–362.

Haugland, R.P. (1992) *Molecular Probes. Handbook of fluorescent probes and research chemicals*. Molecular Probes, Eugene, USA.

Hille, B. (1992) *Ionic channels of excitable membranes*. Sinauer Assoc., Sunderland, MA.

Horn, R. (1991) Estimating the number of channels in patch recordings. *Biophys. J.* **60**, 433–439.

Kolb, H.-A. and Läuger, P. (1978) Spectral analysis of current noise generated by carrier mediated transport. *J. Membr. Biol.* **41**, 167–187.

Läuger, P., Benz, R., Stark, G., Banmberg, E., Jordan, P.C., Fahr, A. and Brock, W. (1981) Relaxation studies of ion transport systems in lipid bilayer membranes. *Quart. Rev. Biophys.* **14**, 513–598.

Maack, J. (1991) *Temperatur- und spannungsabhängige Stromfluktuationen in planaren Phospholipid-bilayern*. Diplomarbeit, Göttingen, FRG.

Sachs, F., Neil, J. and Barkati, N. (1983) The automated analysis of data from single ionic channels. *Pflügers Arch.* **395**, 331–340.

Sakmann, B. and Neher, E. (1983) *Single-channel recording*. Plenum Press, New York.

CHAPTER EIGHT

# Advanced experiments and examples of data

Up to now we have discussed basic technical and theoretical questions of bilayer experiments. This book, however, has been written not only to support inexperienced and experienced scientists in doing successful bilayer experiments, but also to show what successful bilayer experiments may look like. We therefore conclude by describing a series of interesting bilayer experiments that have been published by several authors.

We will start by briefly describing some types of experiments that have been done with pure lipid membranes, then proceed to the investigation of 'simpler' ion-transport systems which are models for bilayer experiments in general, and then go on to questions of protein (ion-channel) reconstitution into planar lipid bilayers. Finally, we will describe a few less typical types of bilayer experiments that can be used as tools for other investigations, for example, protein purification or membrane fusion.

## 8.1 HOW TO CONTROL THE PHYSICAL PROPERTIES OF PLANAR LIPID BILAYERS

The most simple system used in bilayer experiments is a bilayer made from one pure (synthetic) lipid, a *one* component system. This system is very well defined in its biochemical parameters; however, the physical parameters of such a system may still be variable. One of the most important aspects in bilayer experiments therefore is to control these physical parameters and investigate the bilayer properties dependent on them. This is a prerequisite of experiments done with bilayers with more complicated biochemical structure (multi-component system). Indeed, the control of physical parameters in bilayer experiments is often neglected in reconstitution experiments. The consequence of doing so may be unexpected scatter of the data when investigating membrane transport processes mediated by proteins incorporated into the bilayer. In the following we will describe some examples of how to control some of the important physical parameters of bilayer experiments which are frequently not kept under good experimental control.

### 8.1.1 Bilayers below phase transition temperature

We assumed that the control of temperature in a bilayer is a trivial requirement. However, one should bear in mind that using bilayers made from undefined lipid mixtures may add questions about the physical properties of the different lipids involved. One very important parameter of a lipid is its phase transition temperature (in case the used lipid has a defined phase transition being in the range of the temperature of the experiment). We here want to discuss only the main phase transition of the hydrocarbon chains. Above this phase transition the hydrocarbon chains are fluid and therefore quite flexible, allowing, for example incorporated proteins or polypeptides to move freely (as long as they are not fixed by other structures in a well-defined location). When the hydrocarbon chains are immobilized, however, due to freezing the membrane, transport across it may be affected strongly due to immobilizing the proteins or polypeptides.

To stop things being too complicated, it is better not to use lipid mixtures but instead to do experiments concerning the phase transition with pure lipids that have phase transitions in easy-to-handle and physiologically acceptable temperature ranges. Most of the lipids in natural membranes have their phase transition at temperatures significantly lower than the physiological temperature range of the organism they originate from (around 0°C), however, there are natural lipids that have their phase transition in the range of 20–40°C.

Experiments concerning the phase transition of lipids in bilayers were done, among others, by Boheim and co-workers in the early 1980s (Boheim *et al.*, 1980). They made bilayers from 1,3–stearoyl, myristoyl-glycerophosphatidylcholine (1,3–SMPC). This lipid is not present in natural membranes, but as a model system for planar lipid bilayer experiments it has a few significant advantages:

1. The phase transition temperature is about 29°C, thus in a physiologically interesting and experimentally easy to control range.
2. Membranes from this lipid are stable above and below phase transition temperature; in earlier experiments by other groups it was shown that this is not true for most other lipids below phase transition (see Krasne *et al.*, 1971).
3. It is quite a stable chemical (no double bonds).
4. It is a synthetic lipid and therefore a 'pure' lipid.

From this lipid (of course other lipids with comparable properties can be used) folded bilayers as described in Section 5.2.1 were made and the parameters of the pure lipid membrane, as well as those of some incorporated transport systems, were investigated in the range of the phase transition temperature.

In Figure 8.1 the behaviour of the bilayer capacity and the current across a 1,3–SMPC bilayer are shown as functions of temperature. The temperature was slowly changed from above the phase transition to below it. As can be seen, at the range of the phase transition temperature, current fluctuations across the bilayer are observed at low membrane potential (similar observations were made by other groups (see Antonov *et al.*, 1980), and the total membrane capacity becomes smaller. The current fluctuations were interpreted as instabilities in the bilayer during the phase transition, and may be due to phase boundaries as long as a multi-phase system exists in the phase transition range. The reduced capacity is most probably a consequence of a thickening of the bilayer when the hydrocarbon chains are immobilized.

These experiments seem to be quite simple and in fact the ideas behind them are simple and straightforward, however, the experiment is extremely uninteresting as even the bilayers made from specialized lipids are mechanically very unstable in the phase transition region and always tend to break. Thus when doing such experiments one must have a high tolerance to frustration and spend a lot of time to get sufficient results for statistical analysis!

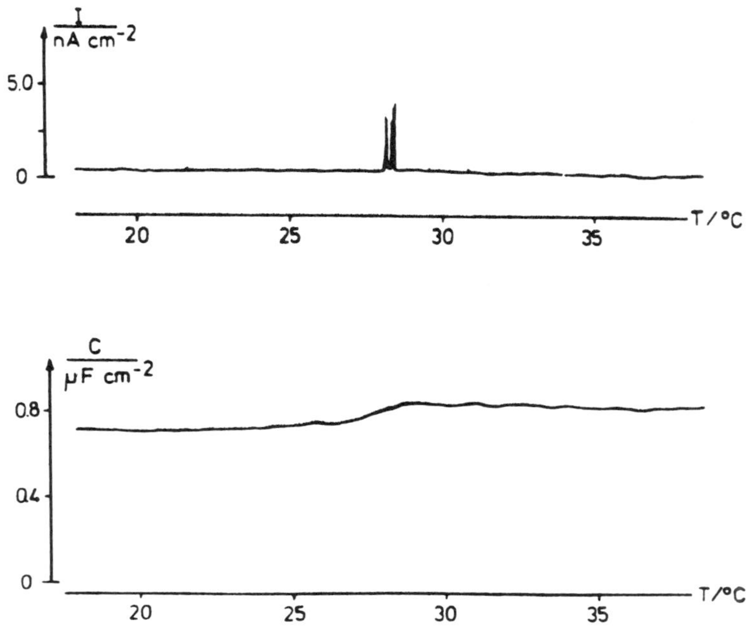


**Figure 8.1** Behaviour of physical membrane parameters in a planar lipid bilayer in the range of the phase transition of the lipid used. A folded bilayer was made from 1–stearoyl-3–myristoyl-glycero phosphatidylcholine (1,3–SMPC) on a Teflon partition with an aperture of about 150 μm diameter. The temperature was significantly higher than the phase transition temperature, $T_c$, of the lipid ($T_c$ = 29°C). Then the bilayer was cooled down slowly to temperatures below $T_c$. In the upper trace, the current across the bilayer is given as a function of temperature (the applied potential was 50 mV), in the lower trace the capacity is shown (measured as discussed in Chapter 3). At the phase transition temperature, spontaneous current fluctuations are to be seen in the current recording, in parallel the bilayer capacity is significantly smaller than at $T_c$ (Reproduced from Boheim *et al.*, 1980, with permission).

In the case of experiments with pure lipids in the phase transition range, it is extremely important to avoid any contamination of the set-up, the aperture, and especially the lipid. Any foreign material in the lipid will of course negate attempts to a achieve a *one* component system and affect the sharpness of the phase transition. Furthermore impurities in the system can be expected to lead to fluctuations in the critical phase transition range. We therefore want to reiterate our comments in Chapter 3 about cleaning the set-up. With respect to the quality of the lipid used, it is necessary to have a pure lipid as far as is biochemically possible.

Another question that is important in this context is the type of bilayer used in experiments in the phase transition range of pure lipids. As shown earlier, painted bilayers still contain a lot of organic solvent when they are black, and so they should not be used with such experiments. Furthermore, the pre-treatment of the partition on which the bilayer is made should be reduced to the absolute minimum. It is tempting to consider using dip-stick bilayers without any pre-treatment from spread liposome monolayers.

As stated, membrane transport systems may be affected strongly by phase transitions in membranes, and this question has been investigated in further experiments (Boheim *et al.*, 1980). At the beginning, simpler model systems were incorporated into bilayers identical to those just described, and the protein-mediated transport within and below the phase transition of the lipid was studied. We here want to discuss

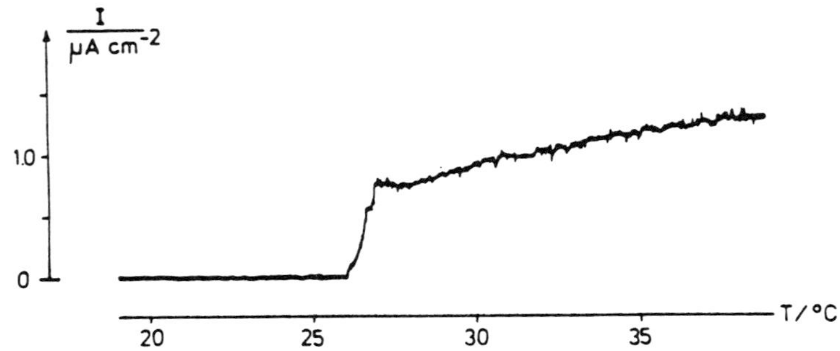


**Figure 8.2** Behaviour of valinomycin-mediated ion transport across a folded bilayer made from 1–hexadecyl-2–tetradecyl-glycero-phosphatidylcholine (1,2–HTPC), the phase transition temperature of this lipid is $T_c$ 27°C. The bilayer was made well above the phase transition temperature, then valinomycin was added to the aqueous solution. In the trace shown, the current across the bilayer is given as a function of temperature. A few degrees below $T_c$ the current across the bilayer (at a small membrane potential) is reduced to the value of a pure lipid membrane, indicating that in the frozen membrane no carrier-mediated ion transport is possible (Reproduced from Boheim *et al.*, 1980, with permission).

briefly a few results for valinomycin, a potassium carrier, and gramicidin, a very well-described pore forming polypeptide.

Valinomycin is a potassium-selective ion carrier, and in order to move ions across a membrane as described in Section 2.3.2.2 it must be able to diffuse freely in the hydrophobic core of the bilayer. Obviously this should not be any problem in a bilayer above phase transition temperature. If the bilayer is frozen, however, it might be expected that valinomycin could no longer diffuse freely in the membrane, thus ion transport across the bilayer should be strongly reduced. In Figure 8.2 it is demonstrated how the current across a bilayer changes when a valinomycin-treated bilayer is cooled down below the phase transition temperature. The current across the bilayer was measured as a function of temperature at a low potential. Valinomycin was added to the aqueous phases after a folded bilayer was made, well above phase transition temperature, and then the system was cooled down. As can be seen, slightly below the phase transition temperature, the current across the bilayer became very low, indicating that carrier mediated-transport was no longer possible, just as was expected from our discussion.

Experiments comparable to these done with valinomycin can be done with any other ion transport system. Gramicidin is a pore-forming polypeptide, which has been investigated in planar lipid bilayers in detail. The structure and the pore forming-mechanism of gramicidin are shown in Figure 8.3. One gramicidin molecule forms a helix spanning about one monolayer of a bilayer. If gramicidin molecules are present in both monolayers of a bilayer, they can meet each other by lateral diffusion and build aqueous pores by dimerization. Through the core of the helices built by dimerisation an aqueous pathway across the membrane is formed as shown in the Figure. This mechanism requires a free lateral diffusion of the gramicidin monomers. In a frozen bilayer such free diffusion most probably is significantly reduced, and in Figure 8.4 what happens with a gramicidin-treated bilayer is shown. The protocol of the experiment is comparable to the valinomycin experiment and the multichannel-current is shown in the upper part of Figure 8.4. This current is reduced significantly at a few degrees below the phase transition temperature; however, a small current remains.

A few things are obvious from such an experiment. First, the fact that the transport across the bilayer is reduced at temperatures lower than the phase transition temperature can only be explained by phase separation in a two-component system. Second, the ion transport

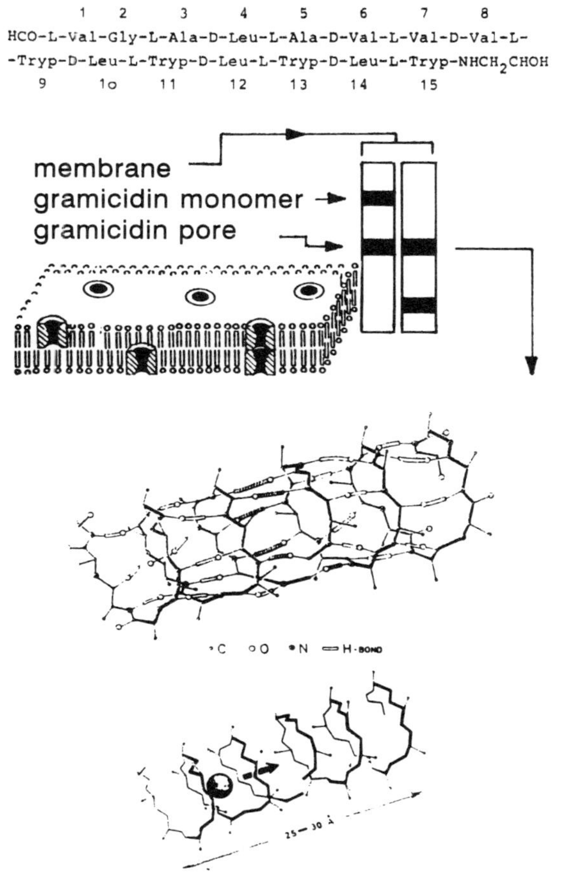


**Figure 8.3** Molecular structure and pore-forming mechanism of gramicidin. Each molecule forms a helix which spans about one monolayer. The helices can freely diffuse laterally in the monolayers of the bilayer and if they are present in both monolayers they may meet and dimerize. Thus a helix spanning the entire bilayer is formed. The ion channel is formed by the interior core of this helix. The lifetime of single pores is defined by dimerization and dissociation of the gramicidin monomers.

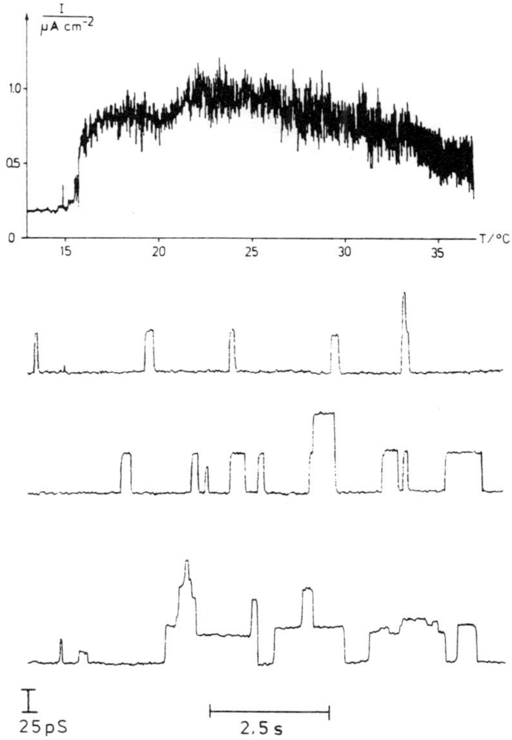


**Figure 8.4** Behaviour of gramicidin pores in bilayers in the range of the phase transition temperature of the lipid. In the upper part of the Figure, the behaviour of a multi-channel system is shown in a 1,3–SMPC bilayer (the principles of the experiment are identical to those given in Figures 8.1 and 8.2). When cooling down the bilayer, first the current mediated by gramicidin increases, and then some degrees below $T_c$ it falls down to a very small value which is, however, significantly bigger than the plain bilayer conductance (Reproduced from Boheim *et al.*, 1980, with permission). In the lower part of the Figure, three single pore traces are shown at temperatures above (first trace) and below (second and third trace) the phase transition temperature of the lipid. Also well below $T_c$, normal gramicidin pores can be observed, however, at a significantly slower kinetics than at above $T_c$. (Reproduced from Hanke, 1981, with permission).

across the bilayer, mediated by gramicidin, really is reduced in the frozen state of the membrane. Finally, the remaining current in an immobilized system may be explained by the freezing of single ion pores in the open state. To give some evidence for this argument, Figure 8.4 (lower part) shows single gramicidin pore fluctuations at different temperatures. It is obvious that the pore kinetics are reduced massively at temperatures below phase transition; however, the pores are not totally immobilized in the given example. Bearing in mind that the gramicidin pore kinetics are slow in any case, and pore lifetimes can be in the range of seconds in fluid bilayers, a real freezing of gramicidin pores is experimentally difficult to verify.

In experiments with other pore-forming polypeptides with much faster kinetics it has been shown, however, that a real freezing of open

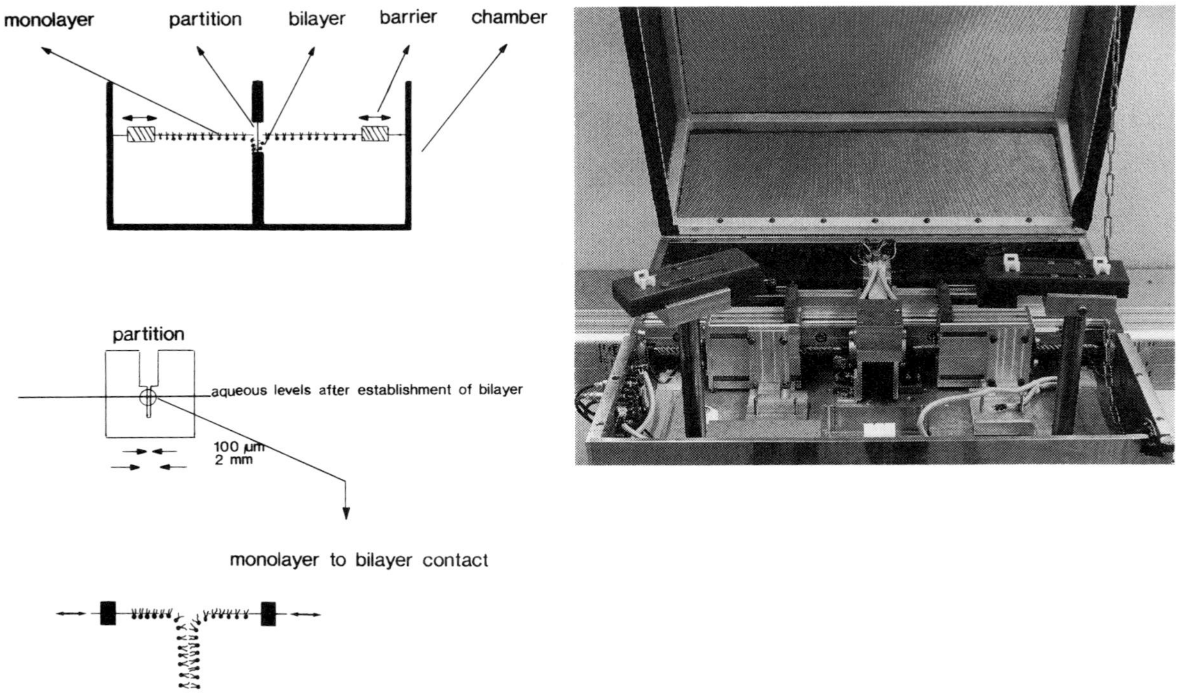


**Figure 8.5** Left: set-up to make bilayers from monolayers on top of two film balances. It is important to combine the two monolayers on a partition with an aperture as shown below, as by this a direct connection between the monolayers and the bilayer is formed without any mechanical barrier (see Dambacher and Fromherz, 1986). Right: photo of a set-up built by the authors (see Chapter 3 for comparison).

pores is possible (in fact the kinetics of alamethicin pores, usually in the millisecond range can be slowed down by at least four to five orders of magnitude by cooling a bilayer below its phase transition temperature).

### 8.1.2 Bilayers formed from monolayers with controlled surface pressure

As discussed already in Section 5.2, the formation of folded bilayers can in principle be done from monolayers of controlled area and surface pressure. The principle of such an experiment is to make the bilayer from monolayers on the air–water interfaces of two film balances, which will give the requested control of the monolayers. In Figure 8.5 a diagram and some detailed information of a set-up which can be used in such experiments is given. This set-up, especially the design of the membrane-carrying aperture in the partition, has the advantage that there is a direct connection between monolayers and bilayer without any boundary that might affect the coupling of monolayers to the bilayer. The principal idea of such a free monolayer-bi-monolayer contact was introduced by Fromherz and co-workers (1986). Another set-up, already shown in Figure 5.14, was introduced by Schindler and co-workers (Schürholz and Schindler, 1983) and makes use of bilayers on tips of glass pipettes. This has, however, the disadvantage of a glass boundary between monolayers and bilayer. It is obvious that an interesting experiment would be to change the lateral monolayer pressure during an experiment with a preformed bilayer. The set-up and experiments are based on the ideas of K. Kaufmann in Göttingen (Kaufmann, 1989). This is shown in Figure 8.6 with a bilayer treated with alamethicin (Boheim, 1974), another pore-forming polypeptide (see Chapter 2). As can be seen, the kinetics of the ion-channel current fluctuations can be changed greatly by changing the monolayer parameters. When using appropriate lipids, the film balances can even be used to produce phase transitions in the monolayers and the properties

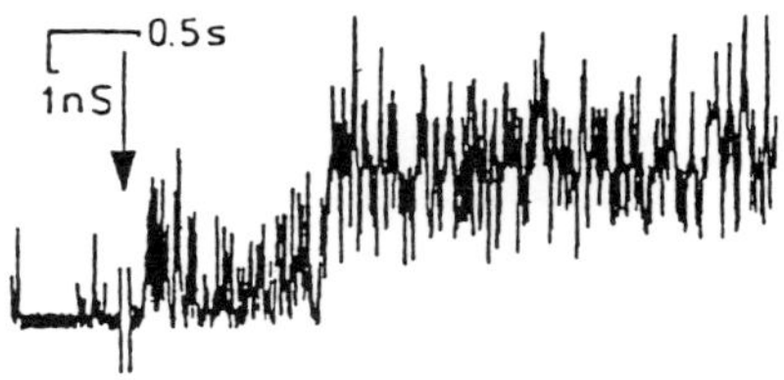


**Figure 8.6** Into a bilayer (made from Asolectin) on a set-up as shown in Figure 8.5 alamethicin was incorporated and a potential was applied to visualize single-pore fluctuations. At the arrow, the areas of the monolayers were reduced thus raising their lateral pressure. As can be seen, the pore activity increases and the pore kinetics slow down.

of the bilayer can be investigated with respect to these in experiments comparable to those for investigation of phase transitions as described in Section 8.1.1.

As such experiments with monolayer-coupled bilayers again are quite unstable mechanically, not many results have been published to date; however, the dependence of membrane transport on physical membrane parameters can be better studied with such a set-up than in bilayer experiments using classical set-ups. This is due to the fact that most of the important physical parameters of a bilayer experiment can be really controlled here.

### 8.1.3 Membrane curvature

In bilayer experiments it is usually assumed that the membranes are flat and that their geometry does not change during an experiment. Sometimes this is controlled during the experiment by electrical capacity measurements. This geometrical stability does not necessarily hold for natural membranes (and sometimes it is not true for bilayers either), and thus an interesting question is to investigate the dependence of physiological membrane properties on parameters like membrane curvature. The introduction of the technique to make bilayers on tips of glass pipettes gives an opportunity to do such experiments. When using a standard patch-clamp pipette holder, suction can be applied to the interior of the pipette (see Chapter 5). By this, a bilayer formed on the tip of a glass pipette, can be changed in its curvature and area as shown in

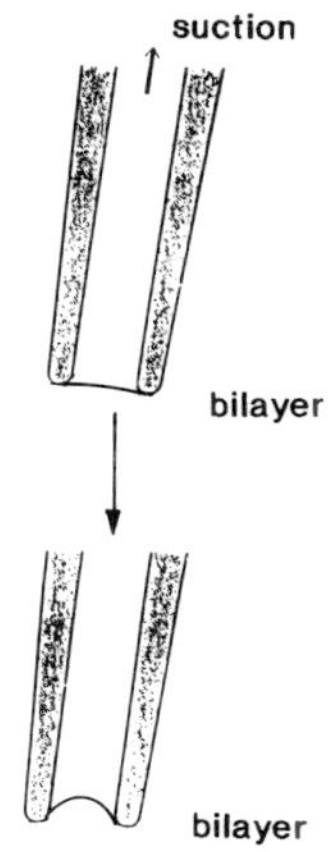


**Figure 8.7** Bilayers on the tip of a glass micropipette using a suction holder. The bilayers can be made from Asolectin for example. In the upper part no suction is applied to the pipette interior, in the lower part the same bilayer is shown when slight suction is applied to it. As is shown, the bilayer has been sucked up into the pipette, and its area and curvature have changed.

Figure 8.7. Without going into details, we would like to point out that the properties of ion transport in such a bilayer can of course be studied as a function of applied suction and membrane curvature. The applied suction can be measured directly; the membrane curvature can either be traced by a electrical capacity measurement (which is not very reliable here, due to the lack of information on the real geometry of the bilayer), or by observing a bilayer under the microscope (the phase boundary – lipid to water – can be visualized using phase-contrast optics). This observation may be the most accurate method but is more costly in time. A video-camera and a computer-aided digital imaging system may be useful in such experiments to calculate the bilayer area and curvature.

## 8.2 INVESTIGATION OF THE ALAMETHICIN PORE

We introduced alamethicin as a good model-system for investigating ion transport mediated by aqueous pores in Chapter 2. In fact alamethicin has been thoroughly investigated in artificial and natural membranes, especially in earlier years of bilayer research. In this Section, we want to describe some alamethicin experiments as they may be useful for general understanding of how to perform bilayer studies with channel-like transport systems. In addition, they can be used to learn how to handle a bilayer experiment in general.

In principle, alamethicin can be incorporated into a bilayer made according to any technique described in Chapter 5. Usually a bilayer is made (almost any lipid can be used) and alamethicin, because of its amphiphilic character, is simply added to the aqueous phase of the bilayer set-up (in the majority of the experiments this is one only, the *cis* side). From there it diffuses into the bilayer and there forms pores as described in Figures 2.11 and 2.12. In Figure 8.8 an alamethicin single-pore fluctuation in a bilayer is shown under constant conditions. The pore can adopt several electrically distinct states, which are defined by the number of monomers making up the pore-forming aggregate. As we know from the discussion in Chapter 2, each change in the pore state is due to the association or dissociation of an alamethicin monomer to an existing aggregate. As seen in Figure 8.8 long fluctuations between different states are due to the formation of one pore which may stay stable over a bigger number of state transitions (together with the trace, an amplitude histogram is given for convenience in Figure 8.8). With a computer program comparable to that discussed in Chapter 7, such a fluctuation can be investigated in detail and single-pore parameters can be extracted (not shown).

Alamethicin was found to be strongly voltage- and concentration dependent in its pore-forming properties, thus we will show here the dependence of alamethicin parameters mainly as functions of these parameters.

The next step in our presentation is to investigate not the single channel parameters, but a multi-channel system as can be produced by adding a high concentration of alamethicin to the aqueous phase at one side of the bilayer. An I/V-curve as shown in Figure 8.9 (left) will result. It is strongly non-linear with high currents at increasing potentials (with asymmetrical alamethicin addition, the curve will be asymmetrical of course). The linear plot of the I/V-curve can be replotted as a half-log plot as given in at the right side of Figure 8.9, for better interpretation. From the slope of the straight line fitted to this half-log plot, an estimate can be made of the number of charges involved (translocated across the bilayer) in the gating process. Comparable to the voltage dependence, the current across an alamethicin-treated bilayer can be plotted as a function of alamethicin concentration (in the aqueous phase). A typical result is given in Figure 8.10 again as a half-log plot. This time the slope of the straight line fitted to the data gives an estimate of the number of alamethicin monomers involved in pore formation. The numbers found over numerous experiments were 4 to 6 (sometimes even higher).

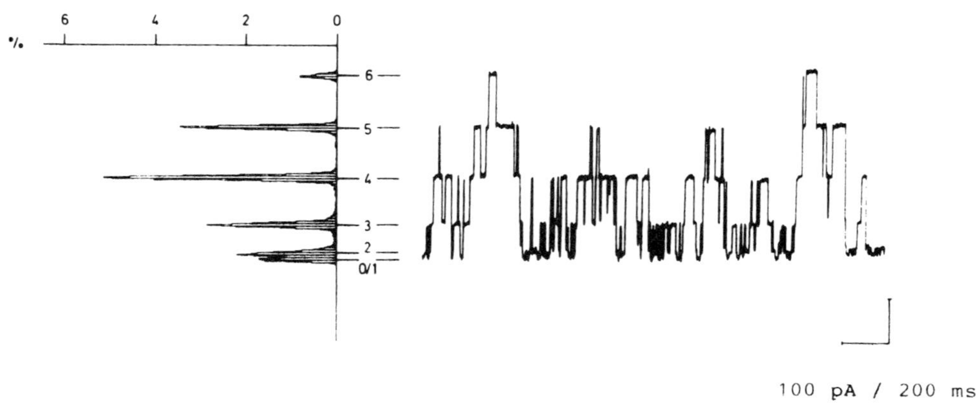


**Figure 8.8** Long trace of single-pore fluctuations induced by alamethicin. A folded bilayer was used. The experiment was done in 1 M KCl at 100 mV, and the drug concentration in the aqueous solution of the *cis* side of the set-up was 50 ng/ml. The whole burst of fluctuations in the trace is due to a single pore event, fluctuating between different conductance levels, the amplitude histogram of a longer fluctuation is shown on the left of the trace (Gelfert-Peuckert, 1986).

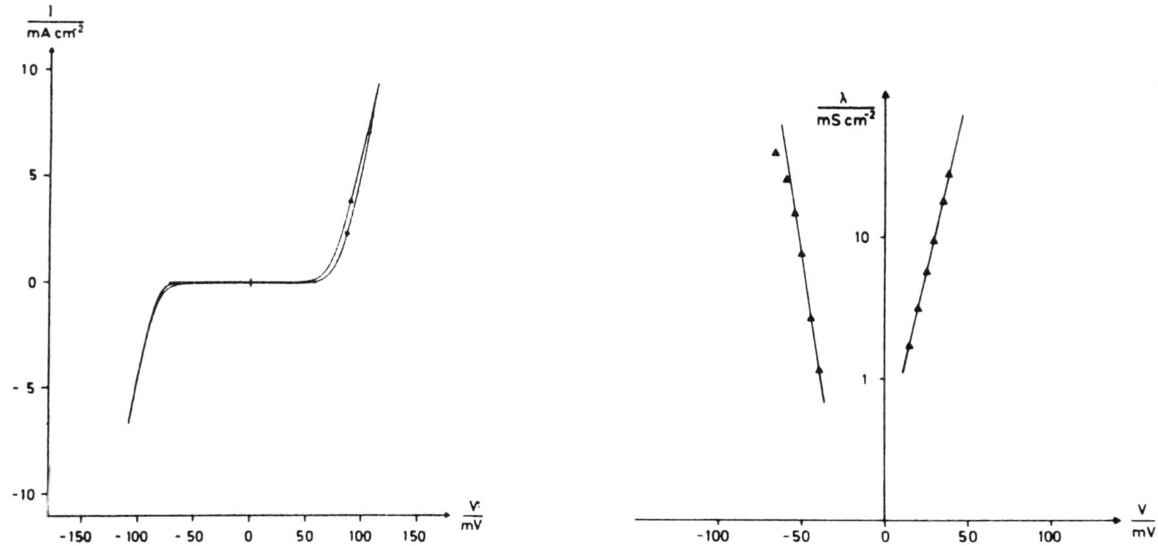


**Figure 8.9** I/V-curve of a bilayer with a high concentration of alamethicin in the bathing solution (0.25 µg/ml). The I/V-curve is strongly nonlinear. At the right, a half-log plot of the bilayer conductance as function of membrane potential is given. Experimental conditions were identical to those in Figure 8.8. The slope of the straight line in the half-log plot can be used to calculate the number of electrical charges involved in pore formation and gating (Gelfert-Peuckert, 1986).

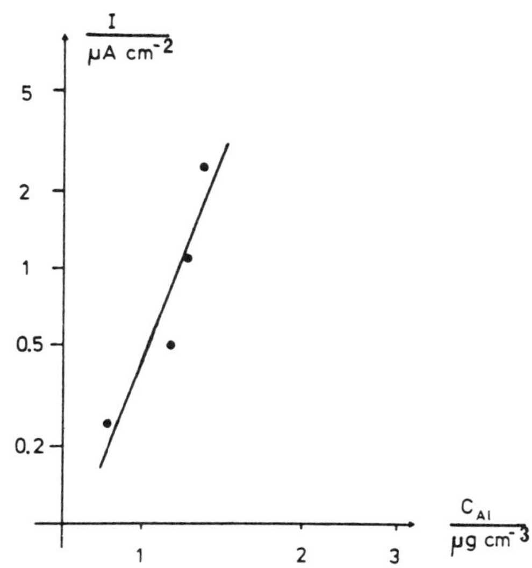


**Figure 8.10** The current across a bilayer is plotted as a function of alamethicin concentration on a half-log scale at a potential of 50 mV in a bilayer made from 1,3–SMPC. The bathing solutions contained 1 M KCl, the temperature was 35°C. The current strongly increases with alamethicin concentration. From the slope of the straight line fitted into the data, an estimate of the number of alamethicin monomers involved in pore formation can be made.

Of course the multi-pore behaviour of an alamethicin-treated bilayer can be explained by the behaviour of single pores. To do this, the single-pore parameters must be measured as functions of potential for example. As found in such experiments, the size, the average pore state and, mainly, the frequency of pore events increases with positive potential. From detailed evaluation of such traces it has been shown that the multi-pore behaviour is mainly defined by the potential dependence of pore formation itself; the number of pore events increases strongly with positive potential as already stated.

We give here a simple procedure for performing an experiment that may give some data about alamethicin.

1. Make a painted bilayer as described in Section 5.1 from pre-purified Asolectin (see Chapter 4).
2. Let the bilayer become black and wait a few minutes to control stability of the bilayer.
3. Add alamethicin to one side of the aqueous solution to a final concentration of 100 ng/ml.
4. Stir for a few minutes and wait for equilibration.
5. Apply +100 mV to the alamethicin addition

side and measure the current across the bilayer; a fluctuation as shown in Figure 8.8 should result.
6. Change the potential and observe changes in the fluctuation pattern.
7. Later in the experiment, one may raise the alamethicin concentration, to see the behaviour of a multi-pore system and to study the concentration dependence.

Alamethicin of acceptable quality for the experiments can be obtained from Sigma chemicals (see Appendix 4).

Up to this point we just have discussed a few very basic things about alamethicin pore formation. This model–system, however, gives the power to learn much more about ion-channel (pore) formation. As shown in Chapter 2, most ion channels of biological membranes are formed by aggregates of α-helical structure. Each alamethicin molecule forms an α-helix in a membrane, too. As alamethicin forms homo-oligomeric complexes that are aggregates of a variable number of α-helices, it can be a good model for structure–function relation studies in ion-channel formation in general. A prerequisite of these studies is the availability of chemically distinct analogues of the alamethicin molecule. Such substances have been synthesized by different groups. Investigating the dependence of pore-forming parameters of such alamethicin analogues on the molecular structure of the drug, may help to understand general features of ion-channel formation and structure–function relations (see Gelfert-Peuckert, 1986; Menistrina *et al.*,1986).

## 8.3 RECONSTITUTION OF THE NICOTINIC ACETYLCHOLINE RECEPTOR CHANNEL

One of the best-known ion channels from biological membranes, maybe the best known of all, is the nAChR-channel. Its molecular structure has been described in Chapter 2. In Chapter 4 we have described general procedures to purify the nAChR channel-forming protein from its natural membrane. Here we now want to describe a few experiments done with nAChR-channels in planar lipid bilayers.

In the last few years much has been learnt about the properties and molecular structure of this channel. Besides peripheral nAChRs from electrocytes and muscle cells which have been investigated at the beginning (Montal *et al.*, 1986), the area of interest has changed during this time to nAChRs found in the nervous system of different species. We will here use the AChR from insect neurons (mainly from *Locusta migratoria*) as a model–system to describe some general properties of ion-channel reconstitution experiments and of nAChR behaviour (Hanke and Breer, 1987).

A prerequisite, of course, is the availability of an appropriate protein preparation. The neuronal nAChR from insects (*Locusta migratoria*) has been purified using an affinity chromatography procedure as described in Section 4.2.2. The protein then was incorporated into lipid-vesicle membranes by dialysis (Section 4.2.4). This preparation was used in the experiments described in the following. The neuronal receptor itself is simpler in its structure than the peripheral receptor, most probably the protein used is a homo-oligomeric aggregate of a 65 kD subunit (see SDS-gel in Figure 4.8).

The starting preparation in our experiments is a 500 mM KCl liposome solution buffered to pH 7.4. The protein is incorporated in the membranes of the liposomes at a low protein to lipid ratio [one or no channel-forming protein molecules in each liposome]. The liposomes are made from Asolectin. This preparation may be used in experiments as follows (any other nAChR preparation may be used of course):

1. Make a folded bilayer from prepurified Asolectin/cholesterol, 90:10 mol% (200 μm diameter) in 200 mM KCl at pH 7.4.
2. Control the bilayer for stability.
3. Make one side of the aqueous solution hyperosmotic by adding 500 mM sucrose and add 1 mM $CaCl_2$ to this side to enhance vesicle fusion.
4. Add 50 μl of the vesicle preparation (containing 0.1–1 μg of protein) to the *cis* side of the bilayer system (which is hyperosmotic).
5. Add 50 μM of carbamylcholine to both sides of the bilayer.

6. Stir and wait for up to 30 min to see whether vesicles fuse with the bilayer.
7. If fusion events occur (as shown in Chapter 6), stop stirring and look for single-channel fluctuation at 50 mV membrane potential.
8. If single-channel fluctuations are present, perfuse the vesicle addition side with aqueous solution identical to the *trans* side to prevent further fusion events.
9. Alternatively further fusion may be stopped by adding 2 mM EGTA to the *cis* side.

Starting from here the properties of the incorporated channels can be investigated with respect to a variety of experimental parameters. In cases when one of the above listed steps fails, stop the experiment and start a new one. Reconstitution experiments may sometimes be frustrating, as the rate of success is not always high. The failure of an experiment in addition may originate from a number of different reasons:

1. the preparation may be denatured
2. the vesicles may not fuse properly with the bilayer
3. the incorporated proteins may be not functional due to inappropriate choice of lipid or another membrane parameter.

These points can not be discriminated between, as the only fact one can observe in the bilayer is the absence of any channel activity; thus additional tests may be necessary in such a case to check for example the quality of the preparation used. Nevertheless, sometimes good results can be obtained from such experiments, as we will now show.

Let us assume a single active AChR-channel has been incorporated into the bilayer successfully, then a current fluctuation recording may look like that shown in Figure 8.11. In a first step, such a fluctuation has to be evaluated in more detail as given in Figure 8.12. Here total amplitude histogram and lifetime histograms of the open and closed state from a long trace identical to Figure 8.11 are shown (the computer program written by the authors and discussed in Chapter 7, was used for this data evaluation; a commercial program can be used instead). As can be seen from the amplitude histogram the channel has electrical distinct open and closed states. From the distance of the two peaks (into these peaks gaussian distribution can be fitted) in the histogram, the conductance of the open state of the channel can be calculated to be 75 pS. The open-state lifetimes are single exponential distributions with a mean lifetime of some milliseconds, the closed state lifetimes are double exponential distributed with lifetimes in the 10 ms range.

By definition, the nAChR is a ligand-activated

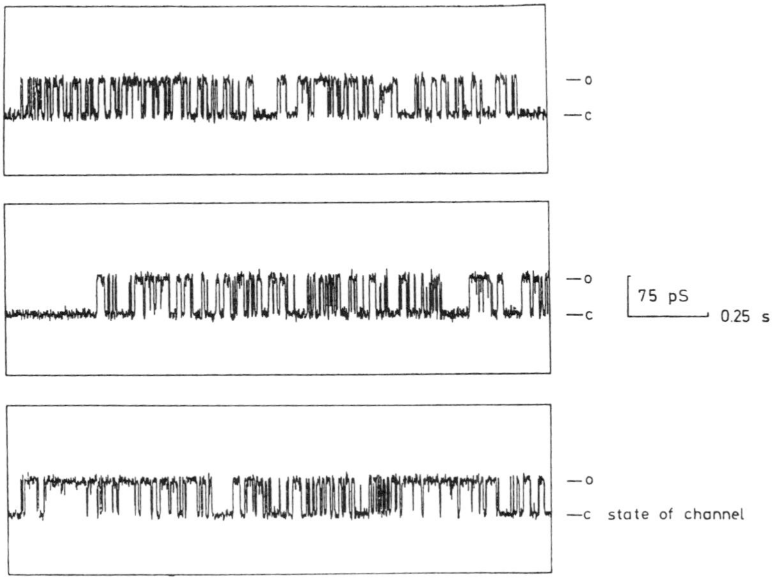


**Figure 8.11** Current fluctuation of a single neuronal nicotinic acetylcholine receptor (nAChR from *Locusta migratoria*) reconstituted into a folded bilayer made from Asolectin. Single-channel events with a well-defined amplitude (channel conductance is 75 pS) can be observed. The mean lifetime of the channel openings is in the millisecond range, the open-state probability is quite high.

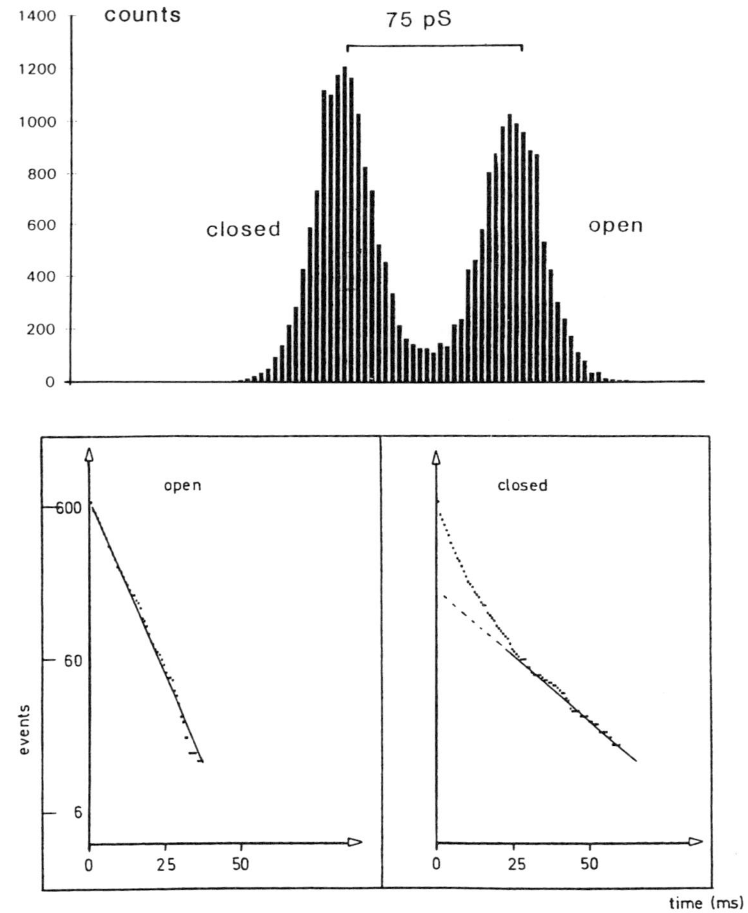


**Figure 8.12** Single-channel parameter evaluation of a channel fluctuation as shown in Figure 8.11. The amplitude histogram of the trace (comparable to Figure 8.11) is shown in the upper part; plots of lifetime of the open and closed channel are shown in the lower part. From the amplitude histogram, a channel conductance of 75 pS can be measured at 50 mV in 200 mM NaCl. The open-state lifetime is a single exponential distribution with an open-state lifetime of about 5 ms, the closed-state lifetime distribution is a double exponential distribution.

channel, thus the next step is to look for the agonist concentration dependence of channel parameters. Experimentally this means that the aqueous solution in the bilayer chamber has to be changed to the desired agonist concentrations. This can be done simply by adding agonist in increasing concentrations to the aqueous solution or, more accurately, by use of a perfusion system as described in Chapter 3. For each agonist concentration an evaluation according to Figures 8.11 and 8.12 is done and the single-channel parameters are plotted as a function of agonist concentration. It is convenient in such a series of experiments, to record in parallel the voltage dependence of the channel parameters at each agonist concentration. The results of such a series of experiments are summarized in Figure 8.13 and Table 8.1. In Figure 8.13 nAChR fluctuation traces are given at different potentials together with an I/V-curve of the open-channel state. From this, the single channel conductance was found to be not significantly voltage dependent; the single-channel kinetics were only slightly voltage dependent.

Open- and closed-state lifetimes as well as the related open state probability (see Chapter 2), however, are strongly agonist concentration dependent. This is more clearly to be seen in Figure 8.14; here the open-state probability is plotted as a function of agonist concentration in a Hill plot together with traces at different agonist concentrations. From the slope of the straight line fitted into the data, a lower limit for the number of agonist molecules necessary to activate the channel can be related; here the slope of the straight line is close to one. Thus, at least one agonist molecule is involved in gating the neuronal nAChR.

**Table 8.1** Properties of the neuronal nicotinic acetylcholine receptor from insects (*Locusta migratoria*).

| | |
|---|---|
| Subunit composition | 1 subunit, 65 kD |
| Protein | Most probably homopentameric, 300 kD |
| Conductance | 75 pS |
| Sublevel | Yes (up to six conductance levels) |
| Agonist cooperativity | ⩾1 |
| Open state lifetimes | In ms range |
| Selectivity | Cationic (not pronounced between cations) |
| pharmacology | Curare blocks |

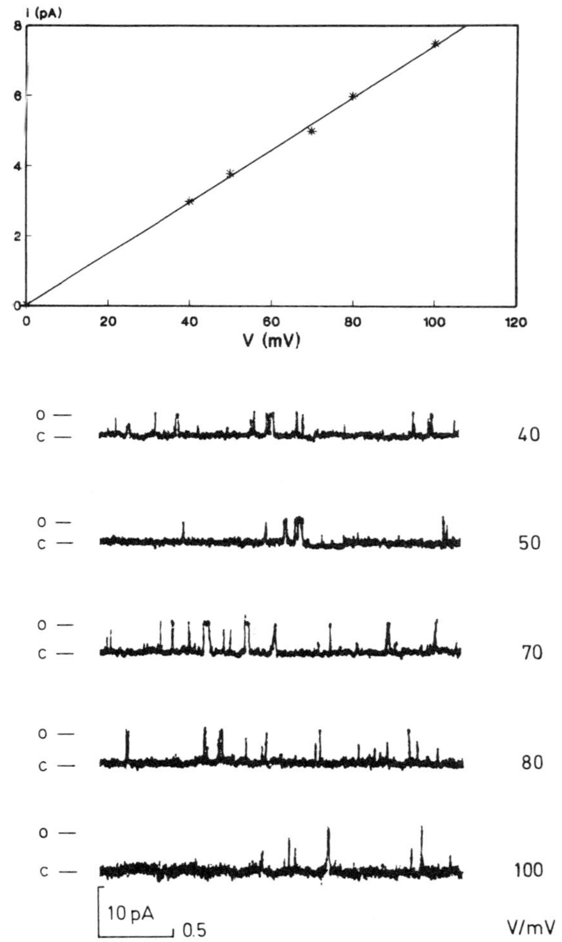


**Figure 8.13** Single-channel fluctuations of neuronal nAChRs as already shown in Figure 8.12 at different membrane potentials. The channel parameters are obviously not significantly potential-dependent. To show this more clearly, together with the traces the I/V-curve of the open-state level is given, which can be fitted by a straight line giving a conductance of 75 pS. As seen from the traces the channel kinetics are not very much potential dependent.

The nAChR is defined not only by its agonist dependence but in addition much is known about the pharmacology of this channel. Pharmacological experiments identical to those changing agonist concentration can be done with reconstituted channels; the desired drug is added to the aqueous solution either directly, or by perfusing the bilayer chambers with aqueous solution of appropriate composition. We will not go into the details of pharmacological experiments here, but do want to point out that a system as described here is unique for studying pharmacological properties of a single purified ion channel, independent of unknown properties of its natural membrane. Nevertheless, it seems to be fair to note that most of our present knowledge about nAChR pharmacology originates from patch-clamp experiments, classical electrophysiology and biochemical studies, and not from bilayer experiments.

After having done a global characterization of the reconstituted nAChR channel, we want to show that each of the channel parameters can be investigated more deeply, using the amplitude distribution as an example here. In Figure 8.15 an amplitude distribution of a single nAChR-channel fluctuation is given at a higher resolution. As can be seen, this distribution again has two main peaks at 0 and 75 pS but it also has some additional activity. The open-state levels between the closed state and the main open state are sometimes called sublevels. Such a distribution can be fitted by a sum of gaussian distributions, and by the position of the peaks different electrically distinct open state levels can be defined as summarized already in Table 8.1. The neuronal nAChR from insects obviously has at least one prounounced sublevel and there may be some less obvious conducting states.

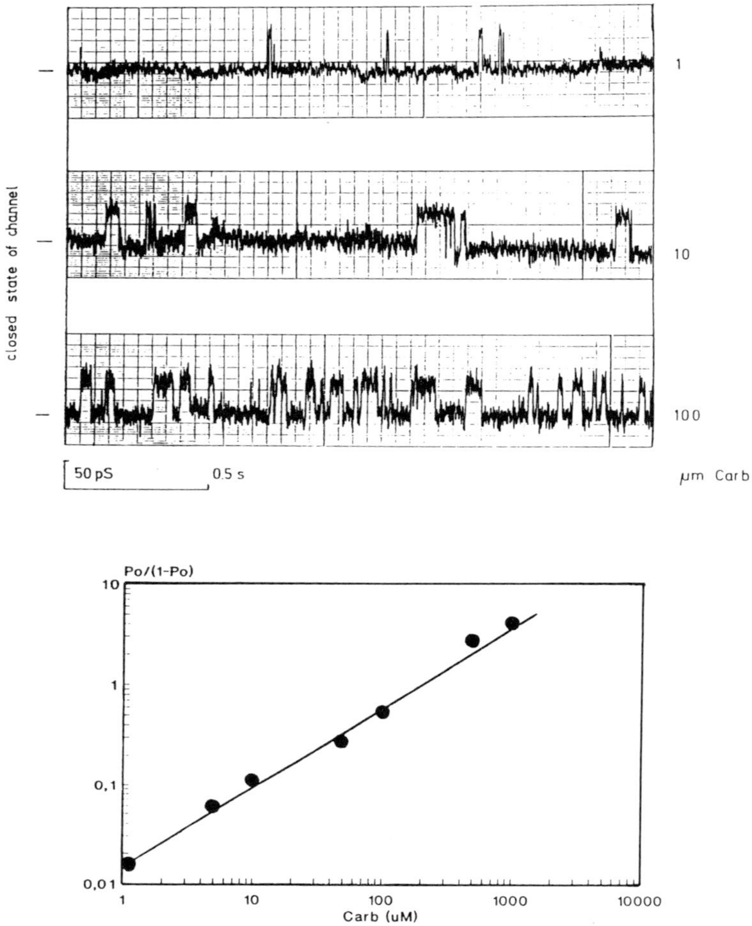


**Figure 8.14** Open-state probability ($P_o$) of the neuronal nAChR reconstituted into a folded bilayer given as a function of agonist concentration. A Hill plot is given together with three traces at different agonist (carbamylcholine) concentration in the aqueous solution. The data can be fitted by a straight line with a slope close to 1.

As stated in the Introduction, one important aim of the studies presented in this book is to examine structure–function relations for functional ion channels. A further step to do this may be to incorporate not the intact protein but the isolated subunit of the neuronal nAChR into a bilayer to investigate the relation of channel parameters to subunit composition. It has been shown that this is possible by using a preparation, as described in Section 4.2.3, made from the neuronal nAChR of insects which is used as a model–system here, as it is composed only of one subunit and therefore is easier to handle in preliminary experiments on this issue (Hanke *et al.*, 1990). Channel fluctuations similar or even identical to those observed for the natural protein were found in planar lipid bilayers after reconstitution of the isolated subunit, indicating that a homo-oligomeric complex of one subunit is able to form a functional nAChR channel different from the more complicated structures of natural peripheral nAChR. Additionally, it was shown by these experiments that denatured subunits of proteins can be renatured and functionally reconstituted. The described technique may be used later in reconstitution of subunits from more complicated proteins, such as the peripheral nAChR with its four subunits.

## 8.4 FURTHER EXAMPLES OF TRANSPORT PROTEINS INVESTIGATED IN PLANAR LIPID BILAYER EXPERIMENTS

In the previous Section we have described the reconstitution of a neuronal nAChR in more

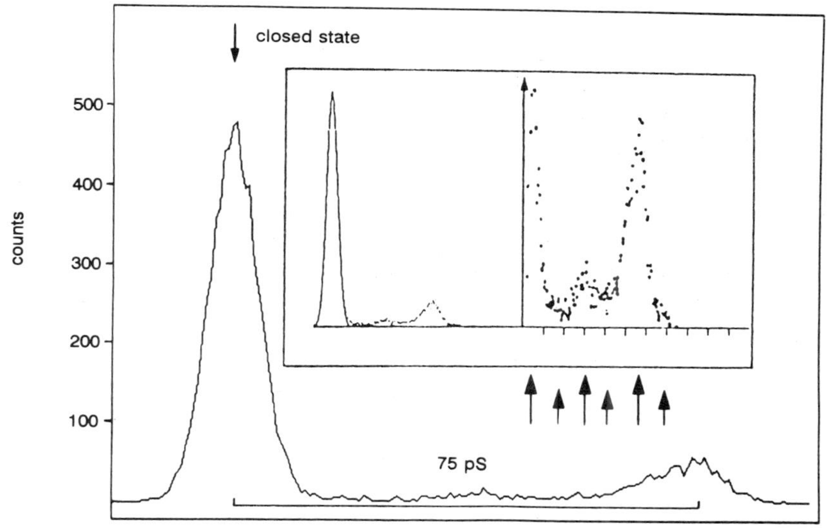


**Figure 8.15** More detailed amplitude histogram of a single-channel fluctuation of a neuronal nAChR reconstituted into a folded bilayer. Besides the two main peaks at 0 pS and 75 pS (see Figure 8.13 for comparison), the graph exhibits a peak at about 40 pS and some additional activity as marked by arrows in the inset, which shows a magnified section of the original graph. Up to six different conductance levels have been found in such graphs.

**Table 8.2** Ion channels that have been successfully reconstituted into planar lipid bilayers (with selected references). The bilayer type given is the one mainly used together with the cited channel for reconstitution. Usually channels can be reconstituted into all types of bilayer, the only exception (studied in detail) is the nicotinic acetylcholine receptor channel, which to date not has been reconstituted into painted bilayers.

| *Channel type* | *Bilayer type* | *Reference* |
|---|---|---|
| Acetylcholine receptor from electrocytes | Folded + dip-stick | Montal *et al.*, 1984<br>Montal *et al.*, 1986 |
| Chloride channel from *Torpedo* electroplax | Folded + painted | Hanke and Miller, 1983<br>Goldberg and Miller, 1991 |
| $Ca^{2+}$-activated $K^+$-channel from muscle | Painted | Moczydlowsky and Latorre, 1983 |
| Sarcoplasmic reticulum potassium channel | Painted | Coronado *et al.*, 1980 |
| Sodium channel from rat brain | Painted + folded | Krueger *et al.*, 1983 |
| Gap junction | Painted | Young *et al.*, 1987 |
| cGMP-activated channel from bovine rod outer segments | Folded | Hanke *et al.*, 1988 |
| $IP_3$-gated ion channel from smooth muscle | Folded | Mayrleitner *et al.*, 1991 |
| $H^+$-channel from chloroplast | Folded | Wagner *et al.*, 1989 |
| Matrix protein | Painted | Benz *et al.*, 1979 |
| DHP-sensitive $Ca^{2+}$-channels | Painted | Mejia-Alvarez *et al.*, 1991 |

* Abbreviations: DHP, dihydropyridine;
$IP_3$, inositol triphosphate

detail. Of course this is just one example out of a large number of other (ion) transport systems that have been reconstituted into planar lipid bilayers. These transport systems can be classified as follows:

1. ionic pores (e.g. alamethicin, gramicidin)
2. ion channels (e.g. nAChR)
3. carriers (e.g. valinomycin)
4. others.

To our knowledge, the main part of recent years' work as far as bilayer experiments are concerned has gone into the reconstitution of ion channels (Hanke, 1985; Miller, 1986); a list of some good examples of reconstituted ion channels is given in Table 8.2 together with references.

Again most of the experiments cited in Table 8.2 are electrophysiological investigations; the use of other techniques together with protein reconstitution into bilayers is still in its infancy.

## 8.5 WHAT CAN BE LEARNED ABOUT PROTEIN STRUCTURE AND FUNCTION FROM BILAYER EXPERIMENTS

Here we return to some statements made in the Introduction about structure–function relations of transport proteins. Obviously, it is important to explore the structure–function relation of transport systems, for example ion channels, as has been done, for example, for the neuronal nAChR, and we have to ask which bilayer experiments can contribute to this. Of course, data from different techniques are needed to obtain some deeper understanding of the function of an ion-channel protein: the structure of any protein can only be explored by biochemical and molecular biological methods.

In the following discussion we want to show what can be done using the insect neuronal nAChR as an example again. From biochemical experiments it has been shown that this channel is composed of one subunit of about 65 kD (Figure 4.8). The intact protein is an aggregate of about 300 kD as was shown by density gradient centrifugation (Hanke *et al.*, 1990). From this and from the fact that other nAChRs are of pentameric structure, it is tempting to state that the functional neuronal nAChR from *Locusta migratoria* as used for our experiments is a homopentameric protein. The structure of the subunit of insect neuronal nAChR has been explored by molecular biological methods and it has been shown that injected mRNA of this single subunit is able to produce functional nAChR channels in oocytes (Marshall *et al.*, 1990). From the fact that five identical subunits make up the functional protein, it may be deduced that the protein can bind up to five ligand molecules and may have distinct open states according to the number of ligand molecules bound [some evidence for that is given from the amplitude histogram shown and from the fact that the sublevel probability was agonist concentration dependent (data not shown)]. Putting all these data together, a gating model as shown in Figure 8.16 can be postulated, having six closed and six open states with a different number of ligand molecules bound (none to five). We want to point out that this evaluation is highly speculative at least with respect to the gating model, but it may indicate what type of questions one can ask of a bilayer experiment, when reference data from other techniques are available.

Comparable ideas can be the basis of many of

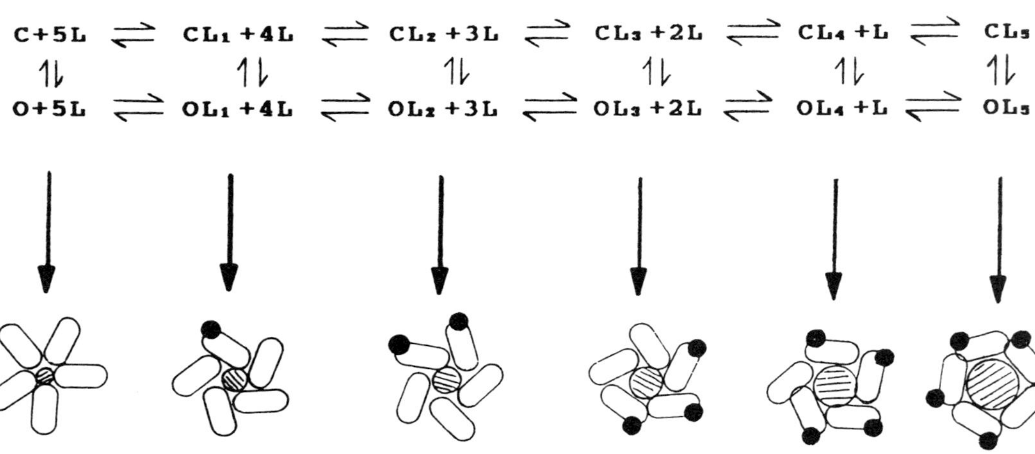


**Figure 8.16** Gating model of a neuronal nAChR (from *Locusta migratoria*). The model was related to the proposed homo-pentameric structure of this protein. It has six closed (C) and six open (O) states, with no or up to five ligand molecules ($L_1$–$L_5$) bound. From the model spontaneous activity without ligand being present, as was found in reconstitution experiments, can be postulated. Each of the open-channel states may have a different conductance (compare with the amplitude graph given in Figure 8.15). In such a case one would expect the sublevel probability to be agonist concentration dependent, as was found in bilayer reconstitution experiments (unpublished data of the authors).

other experiments using planar lipid bilayers and in fact, the number of laboratories working in the field has been increasing in the last few years.

## 8.6 BILAYER EXPERIMENTS AS TOOLS FOR FUNCTIONAL PURIFICATION OF ION-CHANNEL PROTEINS

As we have discussed in Chapter 4, ion-channel proteins can be purified according to one known function of the protein. Of course the most significant function of such a protein is its ability to form a functional ion channel in a membrane. How can this function be used to create an assay for protein purification if no other tool for purification is available? In principle the idea is very simple. The total protein content of a natural membrane is taken (solubilized by detergent) and split into fractions. Column chromatography methods may be used for this as well as gel filtration or other techniques. Each of the fractions then must be tested for its ability to create functional channels having the properties of the channel one is looking for. This test can be done by reconstituting each fraction into a planar lipid bilayer and looking for functional channels. After having found active fractions these may be classified by SDS-gels and must be split into further fractions until finally only one purified protein, as shown on SDS-gels, remains in a fraction giving the desired channel activity in a bilayer reconstitution experiment.

Although the idea seems to be very simple, its realization is extremely time-consuming due to the fact that large numbers of fractions have to be tested carefully in reconstitution experiments. However, some papers have been published showing that such a purification procedure can be successful (e.g. Ide *et al.*, 1991).

## 8.7 OPTICAL MEASUREMENTS ON PLANAR LIPID BILAYERS

Up to now, in our examples we have discussed only electrophysiological measurements and techniques together with planar lipid bilayer experiments. We now want briefly to summarize some optical measurements that may be useful in investigating bilayer properties and properties of incorporated transport systems.

1. In Chapters 3 and 5 we have already shown that a simple microscope is very useful to follow how a painted bilayer becomes black. Using a more accurate system, the area of a black lipid film can be measured (especially when a video-camera and an image-analysis system are coupled to the set-up).
2. It is also useful to trace the fusion of liposomes with bilayers by optical methods as already discussed in Chapter 6. Here liposomes are filled with a dye, which is emptied into the other aqueous solution after fusion and there can be measured (Zimmerberg *et al.*, 1980).
3. Optical methods can be used to trace fluorescence in bilayers when labelled substances are incorporated into the membrane. Let us assume that one component of a liposome membrane has been labelled by a fluorescent dye up to a degree that fluorescence is quenched, then after fusion of the liposome with a bilayer the fluorescence will be dequenched due to diffusion of the dye in the bilayer and can be visualized by a fluorescence microscope (Niles and Cohen, 1991).
4. Fluorescence-labelled lipid in a bilayer can for example be bleached locally with a laser beam and the lateral diffusion of the lipid in the bilayer can be studied by recovery of fluorescence.
5. Bilayers can be made from lipids with triple bonds, which can be polymerized by UV light (Borle *et al.*, 1991).
6. Fluorescent dyes can be used to optically measure true membrane potentials (Cohen and Hoffman, 1982).

## 8.8 STUDY OF MEMBRANE FUSION WITH PLANAR LIPID BILAYERS

The fusion of membranes *in vivo* is a very common process, being important in a lot of physiological situations, such as transmission at

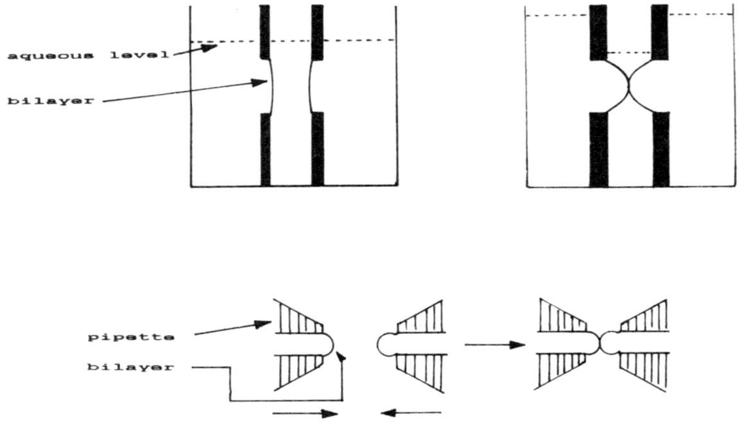


**Figure 8.17** Schematic representation of an experiment to study directly the fusion of two bilayers (upper). Two (painted) bilayers are made on partitions in a distance of a few millimetres (or less) in a three-chamber system and are brought into contact by application of pressure to the outer bilayer chambers. Instead of classical partitions the bilayers can be made even more easily on tips of glass tubes and pressure can be applied to the interior of the tubes by use of appropriate holders (as used in patch-clamp experiments). A sketch of such an experiment is shown in the lower part of the Figure.

chemical synapses, or endocytosis and exocytosis (see earlier). Thus, considerable effort has been directed towards the investigation of membrane fusion in natural and artificial systems. Planar lipid bilayers are perfect systems for studying biophysical properties of membrane fusion due to the free choice of all membrane parameters. Pure lipid membranes can be used, and a single parameter and its effect on the fusion process can be isolated from others. Single fusion events can be produced, and what happens during the fusion event itself can be studied.

We have already met the question of membrane fusion in our discussion about reconstitution of proteins into planar lipid bilayers, by fusion of liposomes with the bilayer (Section 6.2.2). However, membrane fusion in bilayer experiments can be used not only as a tool for reconstitution but as a research aim in itself. Using the techniques described in Sections 6.2.2 and 8.7, for example, the dependence of fusion rates of defined liposomes with a pre-formed bilayer can be studied as functions of a variety of parameters. In addition, other techniques of membrane fusion and studying its properties have been introduced with bilayer experiments (e.g. Wolfe *et al.*, 1991).

One of the most basic experiments is to study the fusion of two defined bilayers directly as has been done by some authors (Kozlov and Markin, 1983). This type of experiment can be done with almost any type of planar lipid bilayer, therefore we will briefly introduce it here. In Figure 8.17 the basic idea of a set-up and an experiment to study the fusion of two bilayers is shown. The two bilayers (painted type) are made on two partitions (glass tubes can be used, too) which are close together, a few millimetres apart. By applying pressure to the opposite chambers or to the interior of the glass tubes, the bilayers expand and approach each other, until they touch. At this moment, the two membranes are very close and one can try to fuse them by using a number of different techniques (for example electrical pulses). The advantage is the possibility to directly study what happens, electrically and under optical control.

## 8.9 LIPID DEPENDENCY OF ION-CHANNEL PARAMETERS

So far we have repeatedly stated that one of the major advantages of planar lipid bilayer reconstitution experiments is the free choice of the lipid composition of the membrane used. This opens the possibility of studying ion-channel parameters as functions of lipid composition in parallel to their dependence on any membrane parameter (Schindler, 1989). Although it is tempting to do such experiments, to date the number of serious investigations of lipid effects on ion transport properties has been extremely low, especially concerning the reconstitution of ion channels (Criado *et al.*, 1984).

More is known about effects of lipid composition on model-systems like gramicidin and alamethicin (Boheim *et al.*, 1983). Another question which has been investigated in more detail in this context is the effect of surface charges (lipids with charged head-groups) on ion transport (e.g. Bell, 1986; Moczydlowski *et al.*, 1985).

Technically, nothing special has to be stated about doing investigations into lipid effects. All the techniques described in this book can be used, and only the composition of the lipid used has to be changed. As an example, one might be interested in changing the amount of cholesterol in a lipid membrane. Here one would just make lipid mixtures with different concentrations of cholesterol and use them together with any of the described techniques.

To answer the question of why not much has been done in this direction is difficult. One of the obvious reasons is the amount of time needed for a series of such experiments. Another reason is the fact that very often channels behave very well in one specific lipid mixture but don't do so in others, so experimenters very often are happy to have found one set of parameters that works and are not willing to take the risk of changing parameters. Finally, as long as one keeps in mind that nature usually does not often change the lipid composition of a membrane after having built it, the question of lipid effects, at least at present, are more interesting for people working in biophysical than those working on physiological questions.

Instead of presenting detailed experiments here, in the reference list we have included some of the few good papers concerning the question of lipid composition in planar lipid bilayer experiments (e.g. McNamee *et al.*, 1986).

## REFERENCES


Antonov, V.F., Petrov, V.V., Molnar, A.A., Predvoditelev, D.A. and Ivanov, A.S. (1980) The appearance of single-ion channels in unmodified lipid bilayer membranes at the phase transition temperature. *Nature* **283**, 585–586.

Bell, J. (1986) The sarcoplasmic reticulum potassium channel: Lipid effects. In: *Ion channel reconstitution*, C. Miller, (ed.), Plenum Press, New York, pp. 469–482.

Benz. R., Janko, K., Boos, W. and Läuger, P. (1979) Ionic selectivity of pores formed by the matrix protein (Porin) of *Escherichia coli. Biochim. Biophys. Acta* **551**, 238–247.

Boheim, G. (1974) Statistical analysis of alamethicin channels in black lipid membranes. *J. Membr. Biol.* **19**, 277–303.

Boheim, G., Hanke, W., and Eibl, H. (1980) Lipid phase transition in planar bilayer membrane and its effect on carrier- and pore-mediated transport. *Proc. Natl. Acad. Sci. USA* **77**, 3403–3407.

Boheim, G., Hanke, W., Überschär, S. and Eibl, H. (1983) Pore forming ionophores induce lateral phase separation in planar lipid bilayers below lipid phase transition. In: *Physical chemistry of transmembrane ion motion*. G. Spach, (ed.), Elsevier Science Publishers, Amsterdam, pp. 11–20.

Borle, F., Sänger, M. and Sigrist, H. (1991) Planar bilayer membranes from photoactivable lipids. *Biochem. Biophys. Acta* **1066**, 144–150.

Cohen, L.B. and Hoffman. J.F. (1982) Optical measurents of membrane potential. In *Techniques in Cellular Physiology*. Elsevier/North-Holland, pp. 1–13.

Coronado, R., Rosenberg, R. and Miller, C. (1980) Ion selctivity, saturation and block in a $K^+$-selective channel from sarcoplasmic reticulum. *J. Gen. Physiol.* **76**, 425–446.

Criado, M., Eibl, H. and Barrantes, F.J. (1984) Functional properties of the acetylcholine receptor incorporated in model lipid membranes. Differential effects of chain length and head-group of phospholipids on receptor affinity and receptor-mediated ion-translocation. *J. Biol. Chem.* **259**, 9188–9198.

Dambacher, K.H. and Fromherz, P. (1986) Reversible bilayer junction of lipid monolayers: free mono-bi-monolayer contact. *Biochim. Biophys. Acta.* **861**, 331–336.

Gelfert-Peukert, S. (1986) *Einfluss des Austauschs und der Modifikation einzelner Aminosäuren des Eicosapeptids Alamethicin auf die Eigenschaften des Ionenkanals*. Dissertation, Bochum, FRG.

Goldberg, F.X. and Miller, C. (1991) Solubilization and functional reconstitution of a chloride channel from *Torpedo californica* electroplax. *J. Membr. Biol.* **124**, 199–206.

Hanke, W. (1981) *Analyse der Eigenschaften porenbildender Antibiotika und Proteine in planaren Lipidmembranen im Bereich der Temperatur der Lipidphasenumwandlung*. Dissertation, Bochum, FRG.

Hanke, W. (1985) Reconstitution of ion channels. *CRC Crit. Rev. Biochem.* **19**, 1–44.

Hanke, W. and Miller, C. (1983) Single chloride channels from *Torpedo* electroplax. Activation by protons. *J. Gen Physiol.* **82**, 25–45.

Hanke, W. and Breer, H. (1987) Characterization of the channel properties of a neuronal acetylcholine receptor reconstituted into planar lipid bilayers. *J. Gen. Physiol.* **90**, 855–879.

Hanke, W., Cook, N.J. and Kaupp, U.B. (1988) cGMP dependent channel protein from photoreceptor membranes: Single channel activity of the purified and reconstituted protein. *Proc. Natl. Acad. Sci. USA* **85**, 94–98.

Hanke, W., Andree, J., Strotmann, J. and Kahle, C. (1990) Functional renaturation of receptor polypeptides eluted from SDS polyacrylamide gels. *Europ. Biophys. J.* **18**, 129–134.

Ide, T., Morita, T., Kawasaki, T., Tagushi, T. and Kasai, M. (1991) Purification of a $K^+$-channel protein of sarcoplasmic reticulum by assaying the channel activity in the planar lipid bilayer system. *Biochem. Biophys. Acta* **1067**, 213–230.

Kaufmann, K. (1989) *Role of lipid membrane*. Caruaru, Brazil.

Kozlow, M.M. and Markin, V.S. (1983) Cell biophysics, possible mechanism of membrane fusion. *Biophysics* **2**, 255–261.

Krasne, G., Eisenman, G. and Szabo, G. (1971) Freezing and melting of lipid bilayers and the mode of nonactin, valinomycin and gramicidin. *Science* **174**, 412–415.

Krueger, B.K., Worley, J.F. and French, R.J. (1983) Single sodium channels from rat brain incorporated into planar lipid bilayer membranes. *Nature* **303**, 172–175.

Marshall, J., Buckingham, S.D., Shingai, R., Lunt, G.G., Goosey, M.W., Darlison, M.G., Sattelle, D.B. and Barnard, E.A. (1990) Sequence and functional expression of a single α subunit of an insect nicotinic acetylcholine receptor. *EMBO J.* **9**, 4391–4398.

Mayrleitner, M., Chadwick, C.C., Timmerman, A.P., Fleischer, S. and Schindler, H. (1991) Purified $IP_3$ receptor from smooth muscle forms an $IP_3$ gated and heparin sensitive $Ca^{2+}$ channel in planar bilayers. *Cell Calcium* **12**, 505–514.

McNamee, M.G., Jones, O.T. and Fong, T.M. (1986) Function of acetylcholine receptors in reconstituted liposomes. In C. Miller, (ed.), *Ion channel reconstitution*. Plenum Press, New York pp. 231–273.

Mejia-Alvarez, R., Fill, M. and Stefani, E. (1991) Voltage-dependent inactivation of T-tubular skeletal calcium channels in planar lipid bilayers. *J. Gen. Physiol.* **97**, 393–412.

Menistrina G., Voges, K.P., Jung. G. and Boheim, G. (1986) Voltage dependent channel formation by rods of helical polypeptides. *J. Membr. Biol.* **93**, 111–132.

Miller, C. (1986) *Ion channel reconstitution*. Plenum Press, New York.

Moczydlowski, E. and Latorre, R. (1983) Gating kinetics of a $Ca^{2+}$-activated $K^+$-channel from rat muscle incorporated into planar lipid bilayers: Evidence for two voltage dependent $Ca^{2+}$-binding reactions. *J. Gen Physiol.* **82**, 511–542.

Moczydlowsky, E., Alvarez, O., Vergara, C. and Latorre, R. (1985) Effect of phospholipid surface charge on the conductance and gating of $Ca^{2+}$-activated $K^+$-channels in planar lipid bilayers. *J. Membr. Biol.* **83**, 273–282.

Montal, M., Labarca, P., Fredkin, D.R., Suarez-Isla, B.A. and Lindstrom, J. (1984) Channel properties of the purified acetylcholine receptor from *Torpedo californica* reconstituted in planar lipid bilayer membranes. *Biophys. J.* **45**, 165–174.

Montal, M., Anholt, R. and Labarca, P. (1986) The reconstituted acetycholine receptor. In: *Ion channel reconstitution*. C. Miller, (ed.), Plenum Press, New York, pp. 157–204.

Niles, W.D. and Cohen F.S. (1991) Fusion of influenza virions with planar lipid membranes detected by video fluorescence microscopy. *J. Gen. Physiol.* **97**, 1101–1119.

Schürholz, T. and Schindler, H. (1983) Formation of lipid–protein bilayers by micropipette-guided contact of two monolayers. *FEBS Lett.* **152**, 187–190.

Schindler, H. (1989) Planar lipid–protein membranes, strategies of formation and of detecting dependencies of ion transport functions on membrane conditions. *Methods Enzymol.* **171**, 225–253.

Wagner, R., Apley, E.C. and Hanke, W. (1989) Single channel currents through reconstituted chloroplast ATP synthase $CF_o - CF_1$. *EMBO J.* **8**, 2827–2834.

Wolfe, J., Bonanno, M. and Chapel, J.-P. (1991) The interaction and fusion of bilayers formed from unsaturated lipids. *Europ. Biophys. J.* **19**, 275–281.

Young, J.D.-E., Cohn, Z.A. and Gilula, N.B. (1987) Functional assembly of gap-junction conductance in lipid bilayers: Demonstration that the major 27 kD proteins forms the junctional channel. *Cell* **48**, 733–743.

Zimmerberg, J., Cohen, F.S. and Finkelstein A. (1980) Fusion of Phospholipid vesicles with planar phospholipid bilayer membranes. Discharge of vesicular content across the planar membrane. *J. Gen. Physiol.* **75**, 241–250.

# Appendices

## A1 LIST OF SYMBOLS AND ABBREVIATIONS USED

| *Symbol* | *Meaning* |
|---|---|
| $\tau$ | Interfacial tension |
| $\varepsilon_r$ | Relative dielectrical constant of a material |
| $\varepsilon_o$ | Absolute dielectrical constant |
| $\mu$ | Chemical potential |
| $\Gamma$ | Concentration |
| $a$ | Activity |
| $A$ | Area |
| AChR | Acetylcholine receptor |
| ATP | Adenosine-triphosphate |
| $C$ | Electrical capacity |
| $C_s$ | Specific electrical capacity |
| $d$ | Distance, membrane thickness |
| $\mathrm{d}_x$ | Derivative of $x$ |
| $e$ | Elementary charge |
| e | Exponential function |
| $E$ | Energy difference |
| $\Delta F$ | Free energy |
| $H$ | Hamaker coefficient |
| $I$ | Current |
| $\Delta P$ | Pressure difference |
| PC | Phosphatidylcholine |
| PE | Phosphatidylethanolamine |
| PI | Phosphatidylinositol |
| PS | Phosphatidylserine |
| $P_o$ | Open state probability of a channel |
| $r$ | Radius |
| $R$ | Gas constant |
| $R$ | Resistance |
| $t$ | Time |
| $T$ | Absolute temperature |
| $T_c$ | Phase transition temperature of a lipid |
| $t_c$ | Closed state lifetime |
| $t_o$ | Open state lifetime |
| $V$ | Potential |

Symbols that are used only once or twice in the text are explained where they are first used. SI units and related units are used throughout the text as far as possible.

## A2 LETTER CODE FOR AMINO ACIDS

| *Amino acid* | *Single letter code* | *Triple letter code* |
|---|---|---|
| Alanine | A | ala |
| Cysteine | C | cys |
| Aspartic acid | D | asp |
| Glutamic acid | E | glu |
| Phenylalanine | F | phe |
| Glycine | G | gly |
| Histidine | H | his |
| Isoleucine | I | ile |
| Lysine | K | lys |
| Leucine | L | leu |
| Methionine | M | met |
| Asparagine | N | asn |
| Proline | P | pro |
| Glutamine | Q | gln |
| Arginine | R | arg |
| Serine | S | ser |
| Threonine | T | thr |
| Valine | V | val |
| Tryptophan | W | trp |
| Tyrosine | Y | tyr |

## A3 LIST OF USEFUL CONSTANTS, DATA AND EQUATIONS

### A3.1 Fundamental constants

| *Constant* | *Symbol* | *Value* |
|---|---|---|
| Absolute electrical constant | $\varepsilon_o$ | $8.854 \times 10^{-12}$ A s $V^{-1}$ $m^{-1}$ |
| Elementary charge | e | $1.602192 \times 10^{-19}$ C |
| Avogadro number | L | $6.0220943 \times 10^{23}$ $mol^{-1}$ |
| Faraday constant | F | $9.64867 \times 10^{4}$ C $val^{-1}$ |
| Gas constant | R | 8.3143 J $K^{-1}$ $mol^{-1}$ |
| Boltzmann constant | k | $1.38062 \times 10^{-23}$ J $K^{-1}$ |

### A3.2 Conversion factors

1 J = 1 N m = 1 Pa m =1 C V = 0.2389 cal
1 J = $6.242 \times 10^{18}$ eV = $2.777 \times 10^{-7}$ kWh
1 atm = 1.01325 bar

### A3.3 Data

#### *A3.3.1 Conductance of selected ion channels*

| *Channel* | *Specificity* | *Mean conductance (pS)* |
|---|---|---|
| Gramicidin | Cations | 10 |
| Alamethicin | None | 1000 |
| Acetylcholine receptor channel | Cations | 50 |
| Glutamate receptor channel | Cations | 120 |
| Glycine receptor channel | Anions | 50 |
| GABA receptor channel | Anions | 20 |
| Sodium channel | $Na^+$ | 10 |
| $Ca^{2+}$-activated potassium channel (SR) | $K^+$ | 180 |
| Calcium channel (L-type) | $Ca^{2+}$ | 25 |
| Chloride channel (*Torpedo*) | $Cl^-$ | 20 |
| Porin | Cations | 2000 |

#### *A3.3.2 Transport rates of some carriers and pumps*

| *Carrier* | *Selectivity* | *Rate (ions/s)* |
|---|---|---|
| Valinomycin | $K^+$ | $4 \times 10^4$ |
| Anion exchanger (erythrocyte) | $Cl^-$ | $5 \times 10^4$ |
| Na/K-pump (kidney) | $Na^+/K^+$ | 150 |

*A3.3.3 Calculated cross-sections of some ion channels*

| *Channel* | *Cross-sectional area (Å²)* |
|---|---|
| Gramicidin | 12.5 |
| Sodium channel (frog neuron) | 15.8 |
| Potassium channel (frog neuron) | 8.6 |
| Porin | 65 |

*A3.3.4 Conductance and capacity of membranes*

| *Membrane type* | *Specific capacity ($\mu F \times cm^{-2}$)* | *Specific conductivity ($ohm \times cm^2$)* |
|---|---|---|
| Painted bilayer | 0.4 | $10^5$–$10^7$ |
| Folded bilayer | 0.8 | $10^5$–$10^7$ |
| Biological membrane | 1.0 | some ohm–$10^6$ |

### A3.4 Equations

Nernst equation:

$$V_m = (RT/zF)\ \ln(c^a/c^i)$$

Goldman equation:

$$V_m = (RT/F)\ln[(P_K c_K + P_{Na} c_{Na} + P_{Cl} c_{Cl})/(P_K c_K + P_{Na} c_{Na} + P_{Cl} c_{Cl})]$$

Michaelis–Menten equation:

$$v = [S]/([S] + K_m)$$

## A4 SOURCES OF SUPPLY

### A4.1 Mechanical parts

NEWPORT CORPORATION, P.O. Box 8020, 18235 Mt. Baldy Circle, Fountain Valley, CA, USA (*tables, positioners, manipulators and other*)

ORIEL CORPORATION, P.O. Box 872, 250 Long Beach Blvd, Stratford, CA, USA (*tables, positioners, manipulators and other*)

SPINDLER & HOYER, Postfach 3353, Königsallee 23, D-3400 Göttingen, FRG (*tables, positioners, manipulators and other*)

BREITENBORN, Postfach 201830, Irmgardstraße 52, D-5600 Wuppertal, FRG, (*Teflon, Teflon film*)

ISEL-AUTOMATION, Im Leibolzgraben 16, D-6419 Eiterfeld 1, FRG (*assorted parts*)

LUIGS & NEUMANN, Boschstrasse 19, D-4030 Ratingen 1, FRG (*positioners, patch-towers and other*)

BURLEIGH INSTRUMENTS, Burleigh Park, Fishers, NY 14453, USA (*manipulators and other*)

### A4.2 Electronics

ANALOG-DEVICES, One Technology Way, P.O. Box 9106, Norwood, MA 02062, USA (*integrated circuits*)

BURR-BROWN, P.O. BOX, 6730 S. Tucson Blvd., Tucson, AZ 85706, USA (*integrated circuits*)

MAXIM INTEGRAL PRODUCTS, 120 San Gabriel Drive, Sunnyvale, CA 94086, USA (*integrated circuits*)

LIST-ELECTRONIC, Pfungstätter Strasse 18–20, D-6100 Darmstadt/Eberstadt, FRG (*amplifiers, filters and other*)

HEKA ELEKTRONIK, Wiesenstrase 71, D-6734 Lambrecht/Pfalz, FRG (*computer operated amplifiers*)

AXON INSTRUMENTS, 1101 Chess Drive, Foster City, CA 94404, USA (*amplifiers, filters and other*)

DAGAN CORPORATION, 2855 Park Avenue, Minneapolis, MN 55407, USA (*amplifiers, filters and other*)

BIO-LOGIC, rue de l'Europe, Z.A. de Font Ratel, 38640 Claix, France (*amplifiers, tape recorders, general equipment*)

WARNER INSTRUMENTS, 1125 Dixwell Avenue, Hamden, CT 06514, USA (*amplifiers and general bilayer equipment*)

WORLD PRECESION INSTRUMENTS, 375 Quinnipiac Avenue, New Haven, CT 06513, USA (*all types of electrophysiological equipment*)

SCIENCE PRODUCTS TRADING, Günthersburgerallee 8, D-6000 Frankfurt, FRG (*all types of electrophysiological equipment*)

CADDOCK, 1717 Chicago Avenue, Riverside, CA 92507–2364, USA (*resistors*)

### A4.3 Data aquisition (hard- and software)

LIST-ELECTRONIC, Pfungstätter Strasse 18–20, D-6100 Darmstadt/Eberstadt, FRG

AXON INSTRUMENTS, 1101 Chess Drive, Foster City, CA 94404, USA

SCIENCE PRODUCTS TRADING, Günthersburgerallee 8, D-6000 Frankfurt, FRG

R.C. ELECTRONICS, 6464 Hollister Avenue, Goleta, CA 93117, USA

### A4.4 Biochemicals

SIGMA, P.O. Box 14508, St. Louis, MO 63178, USA (*general*)

AVANTI POLAR LIPIDS, 2421 Highbluff Rd., Birmingham, AL 35216, USA (*lipids*)

ANALABS, Foxboro Co., 80 Republic Dr., North Haven, CT 06473, USA (*lipids, general*)

FLUKA, Buchs SG, Switzerland (*general*)

MERCK, E., D-6100 Darmstadt, FRG (*general*)

ALDRICH CHEMICALS, P.O. Box 355, Milwaukee, WI 53201, USA (*general*)

MOLECULAR PROBES, P.O. Box 22010, Eugene, OR 97402–0414, USA (*fluorescent dyes*)

### A4.5 Optics

CARL ZEISS, D-7082 Oberkochen, FRG (*microscopes*)

ERNST LEITZ, D-6330 Wetzlar, FRG (*microscopes*)

OLYMPUS OPTICAL, San-Ei Building, 22–2, Nishi Shinjuku, Tokyo, Japan (*microscopes*)

NIKON CORPORATION, Fuji Building, 2–3, Marunuochi 3–chome, Chiyoda-ku, Tokyo, Japan (*microscopes*)

NEWPORT CORPORATION, P.O.Box 8020, 18235 Mt. Baldy Circle, Fountain Valley, CA, USA (*precesion optical parts*)

ORIEL CORPORATION, P.O.Box 872, 250 Long Beach Blvd, Stratford, CA, USA (*precision optical parts*)

SPINDLER & HOYER, Postfach 3353, Königsallee 23, D-3400 Göttingen, FRG (*precision optical parts*)

HAMAMATSU, Arzbergerstr. 10, D-Herrsching, FRG (*video techniques, video imaging*)

SYNOPTICS, 271 Cambridge Park, Milton Road, Cambridge CB4 4WE, UK (*video imaging*)

BRIAN REECE, Turnpike Road, Newbury, Bershire, RG 13 2LW, UK (*video techniques, video imaging*)

DAGE-MTI, 701 N. Roeske Ave, Michigan City, IN, 46360, USA (*video cameras, video techniques*)

# Key references

We have cited specific references on any subject discussed at the end of each chapter. Nevertheless, at the end of the book we would like to add a list of more books and reviews that are useful in working with planar lipid bilayers in general. Some of them already been cited at the end of the chapters, but are repeated here for convenience.

## BOOKS


Alberts, B. (1989) *Molecular biology of the cell.* Garland Pub. New York.

Adam, G., Läuger, P. and Stark, G. (1977) *Physikalische Chemie und Biophysik*. Springer Verlag, Berlin.

Bergelson, L.D. (1980) *Lipid biochemical preparations.* Elsevier, Amsterdam.

Bretscher, M.S. (1985) *The molecules of the cell membrane*. Scientific American Books, New York.

Carfioly, E. and Semenza, G. (eds.) (1979) *Membrane biochemistry: A laboratory manual on transport and bioenergetics*. Springer Verlag, Berlin.

Chad, J. and Wheal, H. (eds.) (1991) *Cellular Neurobiology. A practical approach*. IRL Press, Oxford.

Changeaux, J.-P., Hucho, F., Maelicke, A. and Neumann, E. (eds.) (1985) *Molecular basis of nerve activity*. W. de Gryter, New York.

Conn, P.M. (ed.) (1992) *Electrophysiology and microinjection*. Academic Press, San Diego.

Cook, N.S. (ed.) (1990) *Potassium channels; Structure, classification, function and therapeutic potential.* Ellis Horwood Ltd., New York.

Darnell, J., Lodish, H. and Baltimore, D. (1986) *Molecular cell biology*. Scientific American Books, New York.

Eisenberg, R.S., Frank, M. and Stevens, C.F. (eds.) (1984) *Membranes, channels and noise*. Plenum Press, New York.

Findlay, J.B.C. and Evans, W.H. (ed.) (1987) *Biological membranes: A practical approach*. IRL Press, Oxford.

Hidalgo, C. (ed.) (1988) *Physical properties of biological membranes and their functional implications*. Plenum Press, New York.

Hidalgo, C., Bacigalupo, J., Jaimovich., E. and

Vergara, J. (eds.) (1990) *Transduction in biological systems*. Plenum Press, New York.
Hille, B. (1992) *Ionic channels of excitable membranes*. Sinauer, Sunderland.
Jain, M.K. (1972) *The bimolecular membrane: A system*. Van Nostrand Reinhold, New York.
Jain, M.K. and Wagner (1980) *Introduction to biological membranes*. Wiley, New York.
Kates, M. (1988) *Techniques of lipidology: Isolation, analysis and identification of lipids*. Elsevier, Amsterdam.
Katz, B. (1966) *Nerve, muscle and synapse*. McGraw-Hill Company, New York.
Kruk, Z.L. and Pycok, C.J. (1991) *Neurotransmitters and drugs*. Chapman and Hall, London.
Latorre, R., ed. (1989) *Ionic channels in cells and model membranes*. Plenum Press, New York.
McCormack, J.G. and Cobbold, P.H. (eds.) (1991) *Cellular calcium. A practical approach*. IRL Press, Oxford.
Miller, C. (ed.) (1986) *Ion channel reconstitution*. Plenum Press, New York.
Ostro, M.J. (1983) *Liposomes*. Marcel Dekker, New York.
Plonsey, R. and Barr, R.C. (1988) *Bioelectricity: A quantitative approach*. Plenum Press, New York.
Sakmann, B. and Neher, E. (ed.) (1983) *Single-channel recording*. Plenum Press, New York.
Stamford, J.A., ed. (1992) *Monitoring neuronal activity. A practical approach*. IRL Press, Oxford.
Standen, N.B., Gray, P.T. and Whitaker, M.J. (eds.) (1987) *Microelectrode techniques. The Plymouth Workshop handbook*. Biologists Ltd, Cambridge.
Starzak, M.E. (1984) *The physical chemistry of membranes*. Academic Press, Orlando.
Stein, W.D. (1986) *Transport and diffusion across cell membranes*. Academic Press, Orlando.
Stein, D.W. (1990) *Channels, carriers and pumps*. Academic Press, New York.
Thomas, R.C. (1978) *Ion-sensitive intracellular microelectrodes. How to make and use them*. Academic Press, London.
Ti Tien, H. (1974) *Bilayer lipid membranes*. Marcel Dekker, New York.

## REVIEWS

Alvarez, O., Benos, D. and Latorre, R. (1985) The study of ion channels in planar lipid bilayers. *J. Electrophysiol. Tech.* **12**, 159–177.
Anderson, O.S. and Koeppe R.E. (1992) Model determinants of channel function. *Physiol. Rev.* **72**, 89–157.
Bridges, R.J. and Benos, D.J. (1990) Reconstitution of epithelial channels. *Current Topics Membr. Transp.* **37**, 283–312.
Darzon, A. (1986) Planar bilayers: A powerful tool to study membrane proteins involved in ion transport. *Methods Enzymol.* **127**, 486–502.
Hanke, W. (1985) Reconstitution of ion channels. *CRC Crit Rev. Biochem.* **19**, 1–45.
Hanke, W. and Breer; H. (1989) Reconstitution of acetylcholine receptors into planar lipid bilayers. In: *Subcellular biochemistry*. Vol. 14, J.R. Harris and A.H. Etemandi, (eds.) Plenum Press, New York, pp. 339–362.
Läuger, P., Benz, R., Stark, G., Bamber, E., Jordan, P.C., Fahr, A. and Brock, W. (1981) Relaxation studies of ion transport systems in lipid bilayer membranes. *Quart. Rev. Biophys.* **14**, 513–589.
Lindau, M. (1991) Time resolved capacitance measurements: monitoring exocytosis in single cells. *Quart. Rev. Biophys.* **24**, 75–101.
Schindler, H. (1989) Planar lipid–protein membranes: Strategies of formation and of detecting dependencies of ion transport funtions on membrane conditions. *Methods Enzymol.* **171**, 225–253.
Tamm, L.K. (1991) Membrane insertion and lateral mobility of synthetic amphiphilic signal peptides in lipid model membranes. *Biochem. Biophys. Acta* **1071**, 123–148.
Woolley, G.A. and Wallace, B.A. (1992) Model ion channels: Gramicidin and alamethicin. *J. Membr. Biol.* **129**, 109–139.

# Index